Algebra 2

LARSON
BOSWELL
KANOLD
STIFF

Applications • Equations • Graphs

Chapter 8 Resource Book

The Resource Book contains the wide variety of blackline masters available for Chapter 8. The blacklines are organized by lesson. Included are support materials for the teacher as well as practice, activities, applications, and assessment resources.

McDougal Littell
A HOUGHTON MIFFLIN COMPANY
Evanston, Illinois • Boston • Dallas

Contributing Authors

The authors wish to thank the following individuals for their contributions to the Chapter 8 Resource Book.

Rose Elaine Carbone
José Castro
John Graham
Fr. Chris M. Hamlett
Edward H. Kuhar
Cheryl A. Leech
Ann C. Nagosky
Karen Ostaffe
Leslie Palmer
Ann Larson Quinn, Ph. D.
Chris Thibaudeau

ISBN: 0-618-02016-0

23456789-PBO- 04 03 02 01

Contents

Contents

Contents

Descriptions of Resources

This Chapter Resource Book is organized by lessons within the chapter in order to make your planning easier. The following materials are provided:

Tips for New Teachers These teaching notes provide both new and experienced teachers with useful teaching tips for each lesson, including tips about common errors and inclusion.

Parent Guide for Student Success This guide helps parents contribute to student success by providing an overview of the chapter along with questions and activities for parents and students to work on together.

Prerequisite Skills Review Worked-out examples are provided to review the prerequisite skills highlighted on the Study Guide page at the beginning of the chapter. Additional practice is included with each worked-out example.

Strategies for Reading Mathematics The first page teaches reading strategies to be applied to the current chapter and to later chapters. The second page is a visual glossary of key vocabulary.

Lesson Plans and Lesson Plans for Block Scheduling This planning template helps teachers select the materials they will use to teach each lesson from among the variety of materials available for the lesson. The block-scheduling version provides additional information about pacing.

Warm-Up Exercises and Daily Homework Quiz The warm-ups cover prerequisite skills that help prepare students for a given lesson. The quiz assesses students on the content of the previous lesson. (Transparencies also available)

Activity Support Masters These blackline masters make it easier for students to record their work on selected activities in the Student Edition.

Alternative Lesson Openers An engaging alternative for starting each lesson is provided from among these four types: ***Application, Activity, Graphing Calculator,*** or ***Visual Approach.*** (Color transparencies also available)

Graphing Calculator Activities with Keystrokes Keystrokes for four models of calculators are provided for each Technology Activity in the Student Edition, along with alternative Graphing Calculator Activities to begin selected lessons.

Practice A, B, and C These exercises offer additional practice for the material in each lesson, including application problems. There are three levels of practice for each lesson: A (basic), B (average), and C (advanced).

Contents

Reteaching with Practice These two pages provide additional instruction, worked-out examples, and practice exercises covering the key concepts and vocabulary in each lesson.

Quick Catch-Up for Absent Students This handy form makes it easy for teachers to let students who have been absent know what to do for homework and which activities or examples were covered in class.

Cooperative Learning Activities These enrichment activities apply the math taught in the lesson in an interesting way that lends itself to group work.

Interdisciplinary Applications/Real-Life Applications Students apply the mathematics covered in each lesson to solve an interesting interdisciplinary or real-life problem.

Math and History Applications This worksheet expands upon the Math and History feature in the Student Edition.

Challenge: Skills and Applications Teachers can use these exercises to enrich or extend each lesson.

Quizzes The quizzes can be used to assess student progress on two or three lessons.

Chapter Review Games and Activities This worksheet offers fun practice at the end of the chapter and provides an alternative way to review the chapter content in preparation for the Chapter Test.

Chapter Tests A, B, and C These are tests that cover the most important skills taught in the chapter. There are three levels of test: A (basic), B (average), and C (advanced).

SAT/ACT Chapter Test This test also covers the most important skills taught in the chapter, but questions are in multiple-choice and quantitative-comparison format. (See *Alternative Assessment* for multi-step problems.)

Alternative Assessment with Rubrics and Math Journal A journal exercise has students write about the mathematics in the chapter. A multi-step problem has students apply a variety of skills from the chapter and explain their reasoning. Solutions and a 4-point rubric are included.

Project with Rubric The project allows students to delve more deeply into a problem that applies the mathematics of the chapter. Teacher's notes and a 4-point rubric are included.

Cumulative Review These practice pages help students maintain skills from the current chapter and preceding chapters.

Tips for New Teachers

For use with Chapter 8

LESSON 8.1

COMMON ERROR Students sometimes graph exponential functions of the form $y = a \cdot b^x$ so that the graph touches or even crosses the x-axis. Investigate the *end behavior* of these functions and ask students to determine the *asymptotes* for each function before they connect the points.

COMMON ERROR Students might fail to see the difference between the percent of increase expressed as a decimal, r, and the growth factor, $1 + r$. To show students how these amounts are related but different, write a formula for each:

$$\% \text{ of increase} = r = \frac{\text{New value} - \text{Initial value}}{\text{Initial value}}$$

$$\text{Growth factor} = \frac{\text{New value}}{\text{Initial value}}$$

Then use a concrete example to calculate both amounts.

TEACHING TIP Students can use the Internet, newspapers, or visit a bank to gather information on current interest rates offered by different investments. You might need to explain to your students what some of these investments are, such as CDs, stocks, bonds, mutual funds, etc. Then, you can use the information gathered to create realistic class examples.

LESSON 8.2

TEACHING TIP Have your students compare the graphs of an exponential growth function, $f(x) = a \cdot b^x$, and the *corresponding* decay function, $g(x) = a \cdot \left(\frac{1}{b}\right)^x$. Ask students to compare their domains, ranges, and asymptotes. Students should notice that the graphs of these functions are a reflection of one another in the y-axis.

COMMON ERROR When writing an exponential decay model for the depreciation of an object, some students incorrectly use the formula $y = a \cdot r^t$ rather than $y = a \cdot (1 - r)^t$. Remind students that the *percent of decrease, r,* given in the problem is the value *lost* by the object. Since they use the formula to find the *remaining* value for that object, they need to plug in the *decay factor*, $1 - r$.

LESSON 8.3

TEACHING TIP The following problem will lead students to discover the value of e. Suppose \$1 is invested at 100% annual interest rate for one year. Using the formula for compound interest, we get $A = P \cdot \left(1 + \frac{r}{n}\right)^{nt} = \left(1 + \frac{1}{n}\right)^n$, where the only variable is n, the number of times the money is compounded per year. If the money is only paid at the end of the year, we get $A = \$2$. If the money is compounded biannually, $A = \$2.25$. As n increases, so does A. Can A reach any value we want? Ask students to find A when the money is compounded monthly, weekly, daily, hourly, and so on.

COMMON ERROR Some students think that natural base functions are always exponential *growth* functions, because $e > 1$. Remind students that to decide whether an exponential function is growth or decay, the equation must have only the variable x in the exponent. Therefore, a function such as $f(x) = 5 \cdot e^{-2x}$ must be rewritten as $f(x) = 5 \cdot \left(\frac{1}{e^2}\right)^x$. Since $\frac{1}{e^2} < 1$, the function is an example of exponential decay.

LESSON 8.4

COMMON ERROR Ask students to evaluate $\log_2(-8)$ and you might get any answer from -3, to $\frac{1}{3}$, or just 3. Rewriting logarithmic expressions in exponential form can help students find the correct answer. Suppose $\log_2(-8) = x$. Then $2^x = -8$. Now students can see that there is no answer for this problem, because a power of a positive number can never yield a negative result. This example will help students understand why logarithms are restricted to positive numbers.

TEACHING TIP Students can graph logarithmic functions of the form $y = \log_b(x - h) + k$ using horizontal and vertical translations. To do so, they just need to sketch the graph for $y = \log_b x$ and then move it h units to the right and k units

CHAPTER 8 CONTINUED

Tips for New Teachers

For use with Chapter 8

Lesson 8.5

TEACHING TIP Rather than *giving* students the properties of logarithms or the change-of-base formula, ask them to complete examples until they can come up with the property. For instance, you can ask them to evaluate $\log_3 9^2$ and $2 \cdot \log_3 9$. Then they could evaluate $\log_2 8^3$ and $3 \cdot \log_2 8$. Finally, ask them to use those examples to write an expression equivalent to $\log_b u^n$.

COMMON ERROR Students have a difficult time differentiating $\log_b\left(\frac{m}{n}\right)$ from $\frac{\log_b m}{\log_b n}$ as well as $\log_b(m \cdot n)$ from $\log_b m \cdot \log_b n$. Ask students to evaluate each of these expressions when $b = 2$, $m = 32$ and $n = 4$, to show them that they all yield different results. Then ask students which of these expressions could also be evaluated using the properties of logarithms, and ask students to evaluate them.

COMMON ERROR Watch out for students who *distribute* logarithms, doing things such as $\log_2(x + 5) = \log_2 x + \log_2 5$. Other students who start mixing up multiplication and addition might evaluate $\log_2(x + 5)$ as $\log_2 x \cdot \log_2 5$. Review the properties of logarithms and when to use them to either expand or condense expressions.

Lesson 8.6

TEACHING TIP When solving a logarithmic equation, have students identify the *base* and the *exponent* and then write the equation in exponential form. For instance, if $\log_5(3x + 1) = 2$, then the base is 5 and the exponent is 2, so $5^2 = 3x + 1$.

COMMON ERROR Some students believe that a logarithmic equation cannot have a negative solution, and reject negative values. Give an example such as $\log(x + 7) + \log(x + 4) = 1$. One of the solutions, -2, is a solution because when it is substituted into the equation, there are no logs of negative numbers. The solution -9 is extraneous because $\log(-2)$ and $\log(-5)$ are not defined.

Lesson 8.7

TEACHING TIP All the problems and examples in this lesson specify what type of equation should be used to model the data. Ask students how they would choose either an exponential or a power model if they were not told which to use. Discuss matching the domain and range of the functions to the data or how to use the correlation coefficient, r, and the coefficient of determination, r^2 or R^2.

Lesson 8.8

TEACHING TIP Start the class by discussing how valid it is to use a model that either begins or ends with an *infinite* value. Since in real life few things grow or decrease without bounds, the exponential and power models, as well as other models previously seen, usually have limited validity—their domain needs to be restricted. Logistic models provide a more realistic option, modeling growth within certain bounds for broader domains.

Outside Resources

BOOKS/PERIODICALS

Hurwitz, Marsha. "We Have Liftoff! Introducing the Logarithmic Function." *Mathematics Teacher: Sharing Teaching Ideas* (April 1999); pp. 344–345.

ACTIVITIES/MANIPULATIVES

Goetz, Albert. "Smokey the Bear Takes Algebra." *Mathematics Teacher* (October 1999); pp. 596–600.

SOFTWARE

Masalski, William J. "Topic: Compound Interest." *How to Use the Spreadsheet as a Tool in the Secondary Mathematics Classroom.* NCTM, 1990; pp. 16–19.

VIDEOS

World Population Review. Southern Illinois University at Carbondale, 1990.

NAME ______________________ DATE __________

Parent Guide for Student Success

For use with Chapter 8

Chapter Overview One way that you can help your student succeed in Chapter 8 is by discussing the lesson goals in the chart below. When a lesson is completed, ask your student to interpret the lesson goals for you and to explain how the mathematics of the lesson relates to one of the key applications listed in the chart.

Lesson Title	*Lesson Goals*	*Key Applications*
8.1: Exponential Growth	Graph exponential growth functions and use exponential growth functions to model real-life situations.	• Internet Hosts • Compound Interest • Wind Energy
8.2: Exponential Decay	Graph exponential decay functions and use exponential decay functions to model real-life situations.	• Automobiles • Radioactive Decay • Record Albums
8.3: The Number *e*	Use the number e as the base of exponential functions and use the natural base e in real-life situations.	• Endangered Species • Mount Everest • Rate of Healing
8.4: Logarithmic Functions	Evaluate and graph logarithmic functions.	• Slope of a Beach • Seismology • Tornadoes
8.5: Properties of Logarithms	Use properties of logarithms and apply them to solve real-life problems.	• Acoustics • Photography • Energy in Living Cells
8.6: Solving Exponential and Logarithmic Equations	Solve exponential and logarithmic equations.	• Cooking • History • Oceanography
8.7: Modeling with Exponential and Power Functions	Model data using exponential functions and power functions.	• Communications • United States Stamps • Cities of Argentina
8.8: Logistic Growth Functions	Evaluate and graph logistic growth functions. Use logistic growth functions to model real-life quantities.	• Biology • Owning a VCR • Economics

Study Strategy

Using a Study Group is the study strategy featured in Chapter 8 (see page 464). Encourage your student to get together with other students to form a study group. Help your student find ways to meet with other group members either during the school day or at other times. If more is not feasible, two students can form a study group with telephone discussions.

NAME ______________________ DATE __________

Parent Guide for Student Success

For use with Chapter 8

Key Ideas **Your student can demonstrate understanding of key concepts by working through the following exercises with you.**

Lesson	Exercise
8.1	At the beginning of an experiment, 10,000 bacteria are put in a solution. They grow at a rate of 20% per minute. Write a model giving the population p of the bacteria t minutes after the experiment starts. Use a graph to estimate when there will be 40,000 bacteria in the solution.
8.2	A certain type of rubber ball bounces up to a height of 65% of the height from which it falls. Write a model giving the height h the ball bounces on the nth bounce if it is initially dropped from a height of 5 feet. Use a graph to estimate how many bounces it takes for the ball to first reach a height of less than 1 foot.
8.3	Use a calculator to evaluate $e^{2.1}$.
8.4	Simplify $\log_2 8^x$.
8.5	Condense the expression $3 \log_7 2 + 5 \log_7 x - 4 \log_7 2$.
8.6	You deposit \$1000 in an account that pays 5% interest compounded semiannually. How long will it take for the balance to reach \$1160? (*Hint:* For help in writing an equation to model the situation, see Lesson 8.1. To solve the equation, you can take $\log_{10}$ of each side.)
8.7	A car was worth \$18,000 in 1999 and \$14,400 in 2000. Write an exponential function $y = ab^x$ to fit the data, where x is the number of years after 1998. Use the function to estimate the value of the car in 2002.
8.8	Evaluate $f(0)$ for the function $f(x) = \dfrac{9}{1 + 2e^{-x}}$.

Home Involvement Activity

Directions: Find data on the population of your city or state in 1970, 1980, 1990, and 2000. Decide whether an exponential model fits the points by plotting the natural logarithms of the y-values against the x-values and seeing whether the points $(x, \ln y)$ fit a linear pattern. If appropriate, write an exponential model for the data and use it to predict the population in 2010.

Answers

8.1: $p = 10{,}000(1.2)^t$; about 8 minutes **8.2:** $h = 5(0.65)^n$; 4 bounces **8.3:** about 8.166

8.4: $3x$ **8.5:** $\log_7\left(\dfrac{x^5}{2}\right)$ **8.6:** 3 years **8.7:** $y = 22{,}500(0.8)^x$; \$9216 **8.8:** 3

NAME ______________________ DATE ____________

Prerequisite Skills Review

For use before Chapter 8

EXAMPLE 1 *Evaluating Numerical Expressions*

Evaluate the expression.

a. $\left(\frac{2}{5}\right)^{-2}$ **b.** -4^3

SOLUTION

a. $\left(\frac{2}{5}\right)^{-2} = \frac{2^{-2}}{5^{-2}}$ Power of a quotient property

$= \frac{5^2}{2^2}$ Negative exponent property

$= \frac{25}{4}$ Evaluate powers.

b. $-4^3 = -(4 \cdot 4 \cdot 4) = -64$

Exercises for Example 1

Evaluate the expression.

1. 3^{-3} **2.** $\left(\frac{2}{3}\right)^{-2}$ **3.** -7^{-2} **4.** $\left(\frac{3}{7}\right)^2$

EXAMPLE 2 *Graphing Polynomial Functions*

Describe the end behavior of the graph of the function by completing the statements $f(x) \to$ __?__ as $x \to -\infty$ and $f(x) \to$ __?__ as $x \to +\infty$.

$f(x) = -3x^2$

SOLUTION

To graph the function, make a table of values and plot the corresponding points. Connect the points with a smooth curve and check the end behavior.

x	-3	-2	-1	0	1	2	3
$f(x)$	-27	-12	-3	0	-3	-12	-27

The degree is even and the leading coeficient is negative, so $f(x) \to -\infty$ as $x \to -\infty$ and $f(x) \to -\infty$ as $x \to +\infty$.

NAME ______________________ DATE ________

Prerequisite Skills Review

For use before Chapter 8

Exercises for Example 2

Describe the end behavior of the graph of the function by completing the statements $f(x) \to$ ___?___ $x \to -\infty$ and $f(x) \to$ ___?___ as $x \to +\infty$.

5. $f(x) = x^4$ **6.** $f(x) = -2x^3$ **7.** $f(x) = -5x^2$ **8.** $f(x) = 3x^3$

EXAMPLE 3 *Fitting a Line to Data*

Draw a scatter plot of the data. Then approximate an equation of the best-fitting line.

x	1	2	3	4	5	6	7	8	9	10
y	1.5	1.6	2.1	2.1	2.4	3.1	3.2	3.7	4.2	4.3

SOLUTION

1. Begin by drawing a scatter plot of the data.
2. Next, sketch the line that appears to best fit the data.
3. Then, choose two points on the line. From the scatter plot shown, you might choose (10, 4.3) and (2, 1.6).
4. Finally, find an equation of the line. The line that passes through the two points has a slope of:

$$m = \frac{4.3 - 1.6}{10 - 2} = \frac{2.7}{8} = 0.3375.$$

Use the point-slope form to write the equation.

$y - y_1 = m(x - x_1)$	Use point-slope form.
$y - 1.6 = 0.3375(x - 2)$	Substitute for m, x_1, and y_1.
$y = 0.3375x + 0.925$	Simplify.

Exercise for Example 3

Draw a scatter plot of the data. Then approximate an equation of the best-fitting line.

9.

x	1	2	3	4	5	6	7	8	9	10
y	2.3	2.6	3.3	3.6	4.2	4.4	5.1	5.1	5.7	6.1

NAME ______________________________ DATE ______________

Strategies for Reading Mathematics

For use with Chapter 8

Strategy: Reading Logarithms

In your study of mathematics you have worked with statements such as $2^3 = 8$. One way to interpret this statement is that when you multiply three factors of 2, the result is 8: $2 \cdot 2 \cdot 2 = 8$. Another way to interpret this statement is to say that 3 is the power you raise 2 to in order to get 8. Another way to write this relationship is to use a logarithm:

$\log_2 8 = ?$ *Question:* To what power do you raise 2 in order to get 8?

$\log_{\boxed{2}} 8 = \textcircled{3}$ *Answer:* 2 raised to the **3rd** power equals 8.

exponent

base → $\boxed{2}^{\textcircled{3}} = 8$

In the example above, you saw that $\log_2 8 = 3$ is equivalent to $2^3 = 8$. Any statement of the form $b^k = a$ where $a > 0$, $b > 0$, and $b \neq 1$ can be written in logarithmic form as $\log_b a = k$. This is read as "the log base *b* of *a* is *k*."

EXAMPLE Write each statement in logarithmic form.

a. $2^4 = 16$ **b.** $3^2 = 9$ **c.** $3^{-2} = \frac{1}{9}$ **d.** $25^{1/2} = 5$

SOLUTION

a. $\log_2 16 = 4$ **b.** $\log_3 9 = 2$ **c.** $\log_3 \frac{1}{9} = -2$ **d.** $\log_{25} 5 = \frac{1}{2}$

> **STUDY TIP**
> ***Reading Logarithms***
> When you find a logarithm, you are finding an exponent. To translate between exponential and logarithmic form, first identify the base and the exponent.

Questions

1. In the statement $\log_5 125 = 3$, what is the base? Rewrite this statement in exponential form.
2. Rewrite each statement in logarithmic form.

   a. $2^{-2} = \frac{1}{4}$ **b.** $6^3 = 216$ **c.** $9^2 = 81$ **d.** $16^{1/2} = 4$

3. Rewrite each statement in exponential form.

   a. $\log_{10} 0.001 = -3$ **b.** $\log_7 49 = 2$ **c.** $\log_2 32 = 5$ **d.** $\log_8 2 = \frac{1}{3}$

4. Find each value.

   a. $\log_{10} 100$ **b.** $\log_5 \frac{1}{25}$ **c.** $\log_3 243$ **d.** $\log_{64} 8$

NAME ______________________ DATE __________

Strategies for Reading Mathematics

For use with Chapter 8

Visual Glossary

The Study Guide on page 464 lists the key vocabulary for Chapter 8 as well as reviews vocabulary from previous chapters. Use the page references on page 464 or the Glossary in the textbook to review key terms from prior chapters. Use the visual glossary below to help you understand some of the key vocabulary in Chapter 8. You may want to copy these diagrams into your notebook and refer to them as you complete the chapter.

GLOSSARY

exponential function (p. 465) A function that involves the expression b^x where the base b is a positive number other than 1.

exponential growth function (p. 466) A function of the form $f(x) = ab^x$ where $a > 0$ and $b > 1$.

growth factor (p. 467) The quantity $1 + r$ in the exponential growth model $y = a(1 + r)^t$, where a is the initial amount and r is the percent increase expressed as a decimal..

exponential decay function (p. 474) A function of the form $f(x) = ab^x$ where $a > 0$ and $0 < b < 1$.

decay factor (p. 476) The quantity $1 - r$ in the exponential decay model $y = a(1 - r)^t$, where a is the initial amount and r is the percent decrease expressed as a decimal.

Graphing an Exponential Growth Function

One example of an exponential growth function is compound interest. If you invest \$1000 at 6% interest compounded quarterly, after t years your account will have a balance of $y = 1000(1.015)^{4t}$.

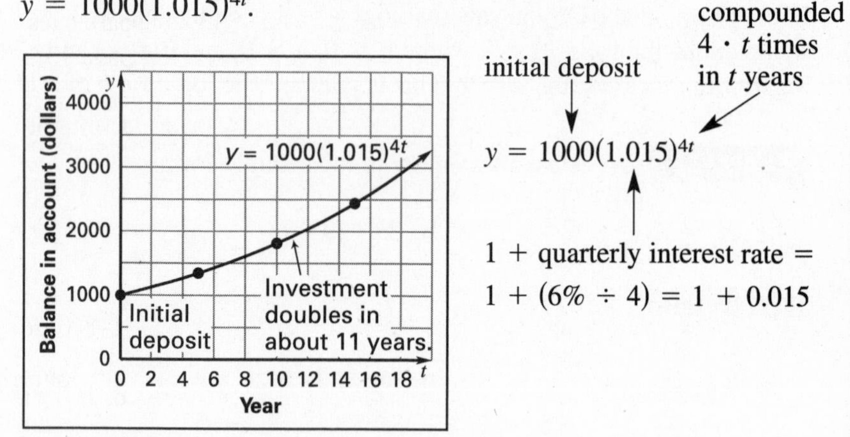

Graphing an Exponential Decay Function

Radioactive decay of an isotope can be represented by an exponential decay function. If a scientist has 1 gram of phosphorus-32 on day 0, the amount remaining d days later is given by the function $y = (0.5)^{d/14}$.

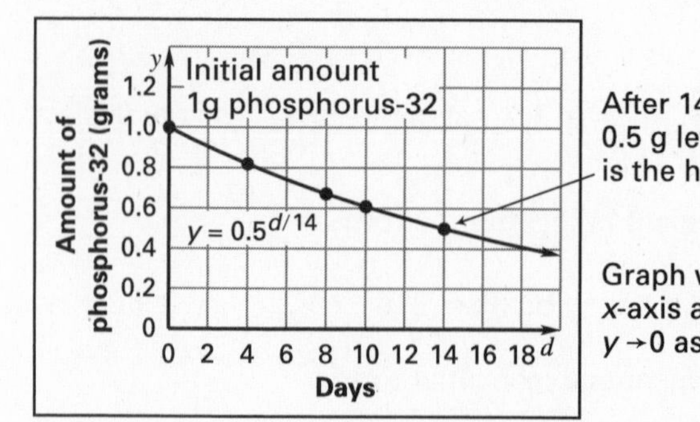

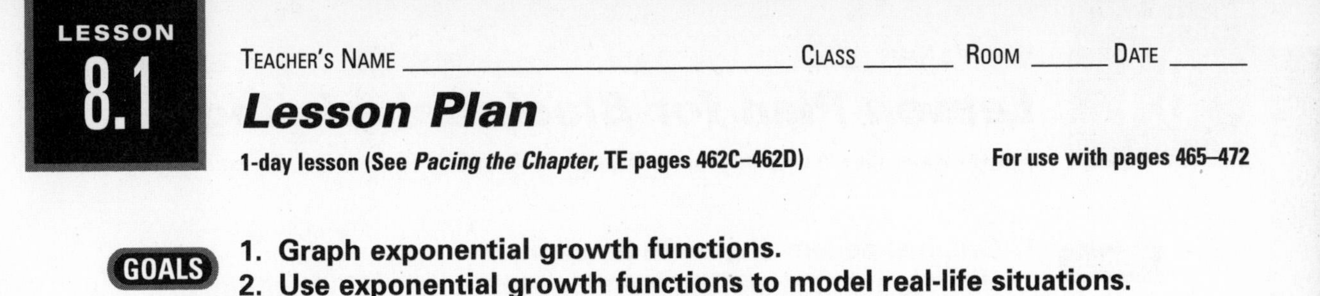

LESSON 8.1

TEACHER'S NAME ______ CLASS ______ ROOM ______ DATE ______

Lesson Plan

1-day lesson (See *Pacing the Chapter,* TE pages 462C–462D) **For use with pages 465–472**

GOALS
1. **Graph exponential growth functions.**
2. **Use exponential growth functions to model real-life situations.**

State/Local Objectives ______

✓ **Check the items you wish to use for this lesson.**

STARTING OPTIONS
____ Prerequisite Skills Review: CRB pages 5–6
____ Strategies for Reading Mathematics: CRB pages 7–8
____ Warm-Up or Daily Homework Quiz: TE pages 465 and 452, CRB page 11, or Transparencies

TEACHING OPTIONS
____ Lesson Opener (Application): CRB page 12 or Transparencies
____ Graphing Calculator Activity with Keystrokes: CRB page 13
____ Examples 1–4: SE pages 466–468
____ Extra Examples: TE pages 466–468 or Transparencies; Internet
____ Closure Question: TE page 468
____ Guided Practice Exercises: SE page 469

APPLY/HOMEWORK

Homework Assignment
____ Basic 14–30 even, 34–40 even, 43–45, 55, 68–69, 71–85 odd, 100
____ Average 14–40 even, 43–48, 55, 66–70, 71–89 odd, 100
____ Advanced 14–40 even, 43–48, 55, 62–70, 71–99 odd, 100

Reteaching the Lesson
____ Practice Masters: CRB pages 14–16 (Level A, Level B, Level C)
____ Reteaching with Practice: CRB pages 17–18 or Practice Workbook with Examples
____ Personal Student Tutor

Extending the Lesson
____ Applications (Real-Life): CRB page 20
____ Challenge: SE page 472; CRB page 21 or Internet

ASSESSMENT OPTIONS
____ Checkpoint Exercises: TE pages 466–468 or Transparencies
____ Daily Homework Quiz (8.1): TE page 472, CRB page 24, or Transparencies
____ Standardized Test Practice: SE page 472; TE page 472; STP Workbook; Transparencies

Notes ______

LESSON 8.1

TEACHER'S NAME ______________________ CLASS ________ ROOM ______ DATE ______

Lesson Plan for Block Scheduling

Half-day lesson (See *Pacing the Chapter,* TE pages 462C–462D) For use with pages 465–472

GOALS
1. Graph exponential growth functions.
2. Use exponential growth functions to model real-life situations.

State/Local Objectives ______________________

____

____

CHAPTER PACING GUIDE	
Day	**Lesson**
1	**8.1 (all)**; 8.2(all)
2	8.3 (all)
3	8.4 (all)
4	8.5 (all)
5	8.6 (all)
6	8.7 (all); 8.8(all)
7	Review/Assess Ch. 8

✓ Check the items you wish to use for this lesson.

STARTING OPTIONS

____ Prerequisite Skills Review: CRB pages 5–6
____ Strategies for Reading Mathematics: CRB pages 7–8
____ Warm-Up or Daily Homework Quiz: TE pages 465 and 452, CRB page 11, or Transparencies

TEACHING OPTIONS

____ Lesson Opener (Application): CRB page 12 or Transparencies
____ Graphing Calculator Activity with Keystrokes: CRB page 13
____ Examples 1–4: SE pages 466–468
____ Extra Examples: TE pages 466–468 or Transparencies; Internet
____ Closure Question: TE page 468
____ Guided Practice Exercises: SE page 469

APPLY/HOMEWORK

Homework Assignment (See also the assignment for Lesson 8.2.)

____ Block Schedule: 14–40 even, 43–48, 55, 66–70, 71–89 odd, 100

Reteaching the Lesson

____ Practice Masters: CRB pages 14–16 (Level A, Level B, Level C)
____ Reteaching with Practice: CRB pages 17–18 or Practice Workbook with Examples
____ Personal Student Tutor

Extending the Lesson

____ Applications (Real Life): CRB page 20
____ Challenge: SE page 472; CRB page 21 or Internet

ASSESSMENT OPTIONS

____ Checkpoint Exercises: TE pages 466–468 or Transparencies
____ Daily Homework Quiz (8.1): TE page 472, CRB page 24, or Transparencies
____ Standardized Test Practice: SE page 472; TE page 472; STP Workbook; Transparencies

Notes ______________________

____

____

NAME ______________________ DATE __________

Available as a transparency

Warm-Up Exercises

For use before Lesson 8.1, pages 465–472

Describe the end behavior of each graph as $x \to \infty$.

1. $f(x) = 3x^3$
2. $f(x) = -3x^3$
3. $f(x) = \frac{1}{3}x^2$

State the domain and range of each function.

4. $y = x^2$
5. $y = \sqrt{x}$

Daily Homework Quiz

For use after Lesson 7.7, pages 445–454

Use the following data set of prices for bread: \$ 0.59, \$0.75, \$0.79, \$0.99, \$0.99, \$1.09, \$1.19, \$1.19, \$1.19, \$1.25, \$1.39, \$1.49, \$1.55, \$1.79

1. Find the mean, median, and mode of the data set.
2. Find the range and standard deviation of the data set.
3. Draw a box-and-whisker plot of the data set.

NAME ______________________ DATE __________

Available as a transparency

Application Lesson Opener

For use with pages 465–472

According to legend, the inventor of the game of chess requested payment as follows:

My request is a modest one. I humbly request one grain of rice on the first square on the chess board, two grains on the second square, four grains on the third square, and so on. Each square represents twice as many grains as the previous one.

1. The king agreed to the request immediately. Without doing any calculations, do you think this was wise?

2. Complete the table for the first 6 squares.

Square	1	2	3	4	5	6
Number of grains	1	2	4			
Exponential form	2^0	2^1	2^2			

3. Write an expression for the number of grains on square x of the chess board.

4. A chess board has 64 squares. How many grains of rice are to be placed on the 64th square?

5. To get an idea how big this number is, suppose that there are 1,000,000 people in the kingdom and that each grain of rice weighs 0.00008 pound. The rice on the 64th square represents how many pounds of rice per person?

6. Do you think the king was able to fulfill the request?

NAME ______________________ DATE __________

Graphing Calculator Activity Keystrokes

For use with page 471

Keystrokes for Exercise 55

TI-82

2500 ENTER

+ 2nd [ANS] × .04 ENTER

Balance after five years.

ENTER ENTER ENTER ENTER

Compounded quarterly:

2500 ENTER

+ 2nd [ANS] × .01 ENTER

Press ENTER 19 more times to get the balance after five years.

TI-83

2500 ENTER

+ 2nd [ANS] × .04 ENTER

Balance after five years.

ENTER ENTER ENTER ENTER

Compounded quarterly:

2500 ENTER

+ 2nd [ANS] × .01 ENTER

Press ENTER 19 more times to get the balance after five years.

SHARP EL-9600c

2500 ENTER

+ 2ndF [ANS] × .04 ENTER

Balance after five years.

ENTER ENTER ENTER ENTER

Compounded quarterly:

2500 ENTER

+ 2ndF [ANS] × .01 ENTER

Press ENTER 19 more times to get the balance after five years.

CASIO CFX-9850GA PLUS

From the main menu, choose RUN.

2500 EXE

+ SHIFT [ANS] × .04 EXE

Balance after five years.

EXE EXE EXE EXE

Compounded quarterly:

2500 EXE

+ SHIFT [ANS] × .01 EXE

Press EXE 19 more times to get the balance after five years.

LESSON **8.1**

NAME ______________________ DATE ________

Practice A

For use with pages 465–472

Match the function with its graph.

1. $f(x) = 3^x$
2. $f(x) = -3^x$
3. $f(x) = 2(3^x)$
4. $f(x) = \frac{1}{2}(3^x)$
5. $f(x) = -\frac{1}{2}(3^x)$

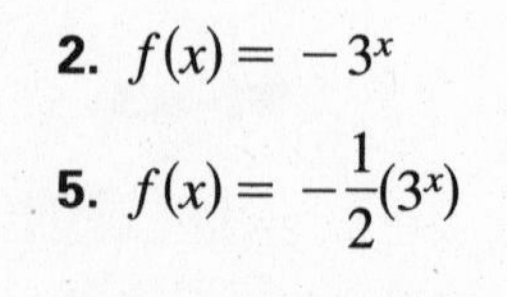

6. $f(x) = -2(3^x)$

A.

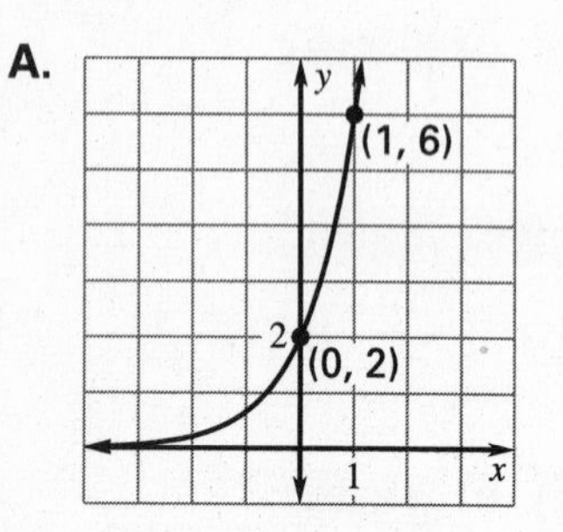

B.

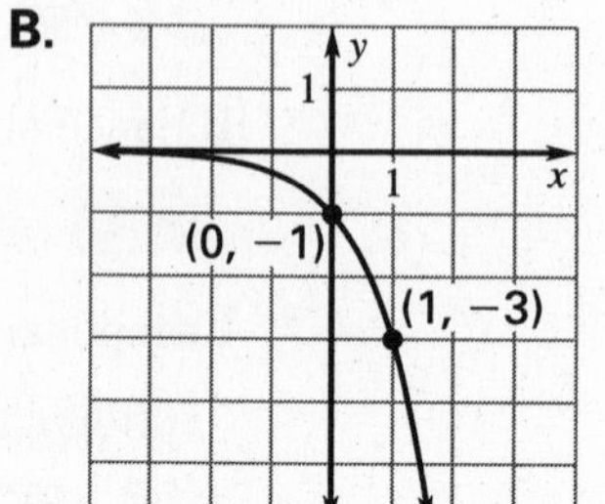

C.

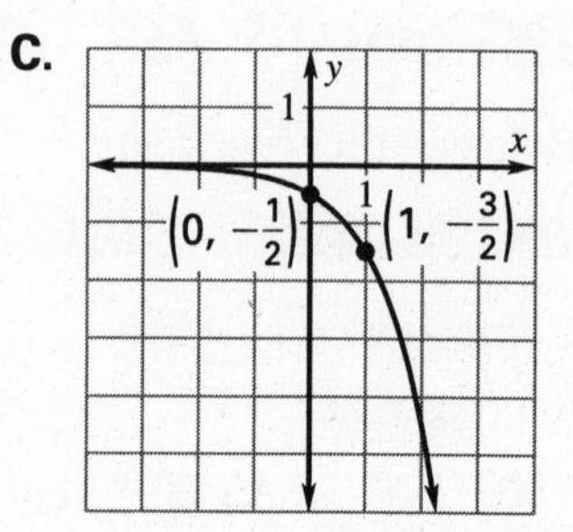

D.

y
1
x
(0, −2)
−2
(1, −6)

E.

y
$(0, \frac{1}{2})$
1
$(1, \frac{3}{2})$
1
x

F.

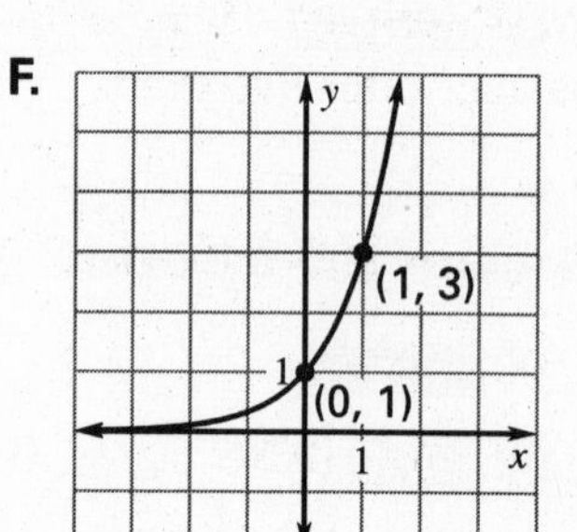

Explain how the graph of *g* can be obtained from the graph of *f*.

7. $f(x) = \left(\frac{4}{3}\right)^x$

   $g(x) = \left(\frac{4}{3}\right)^x + 2$

8. $f(x) = 2^x$

   $g(x) = 2^x - 5$

9. $f(x) = \left(\frac{5}{3}\right)^x$

   $g(x) = \left(\frac{5}{3}\right)^{x+1}$

10. $f(x) = 5^x$

    $g(x) = 5^{x-3}$

11. $f(x) = 2^x$

    $g(x) = -2^x$

12. $f(x) = 3(2^x)$

    $g(x) = 3(2^x) + 2$

Identify the *y*-intercept and asymptote of the graph of the function.

13. $y = 3^x$
14. $y = \left(\frac{6}{5}\right)^x$
15. $y = 2(4^x)$
16. $y = \frac{1}{2}(4^x)$
17. $y = -2(4^x)$
18. $y = -\frac{1}{4}(4^x)$
19. ***Account Balance*** You deposit \$2000 in an account that earns 5% annual interest. Find the balance after 1 year if the interest is compounded with the given frequency.

    a. annually

    b. quarterly

    c. monthly

LESSON 8.1

NAME ______________________ DATE ________

Practice B

For use with pages 465–472

Match the function with its graph.

1. $f(x) = \left(\frac{4}{3}\right)^x + 2$
2. $f(x) = \left(\frac{4}{3}\right)^{x+1} + 2$
3. $f(x) = \left(\frac{4}{3}\right)^{x+2} - 1$
4. $f(x) = 3^{x-1}$
5. $f(x) = -3^{x-1}$
6. $f(x) = -3^{x+1} + 2$

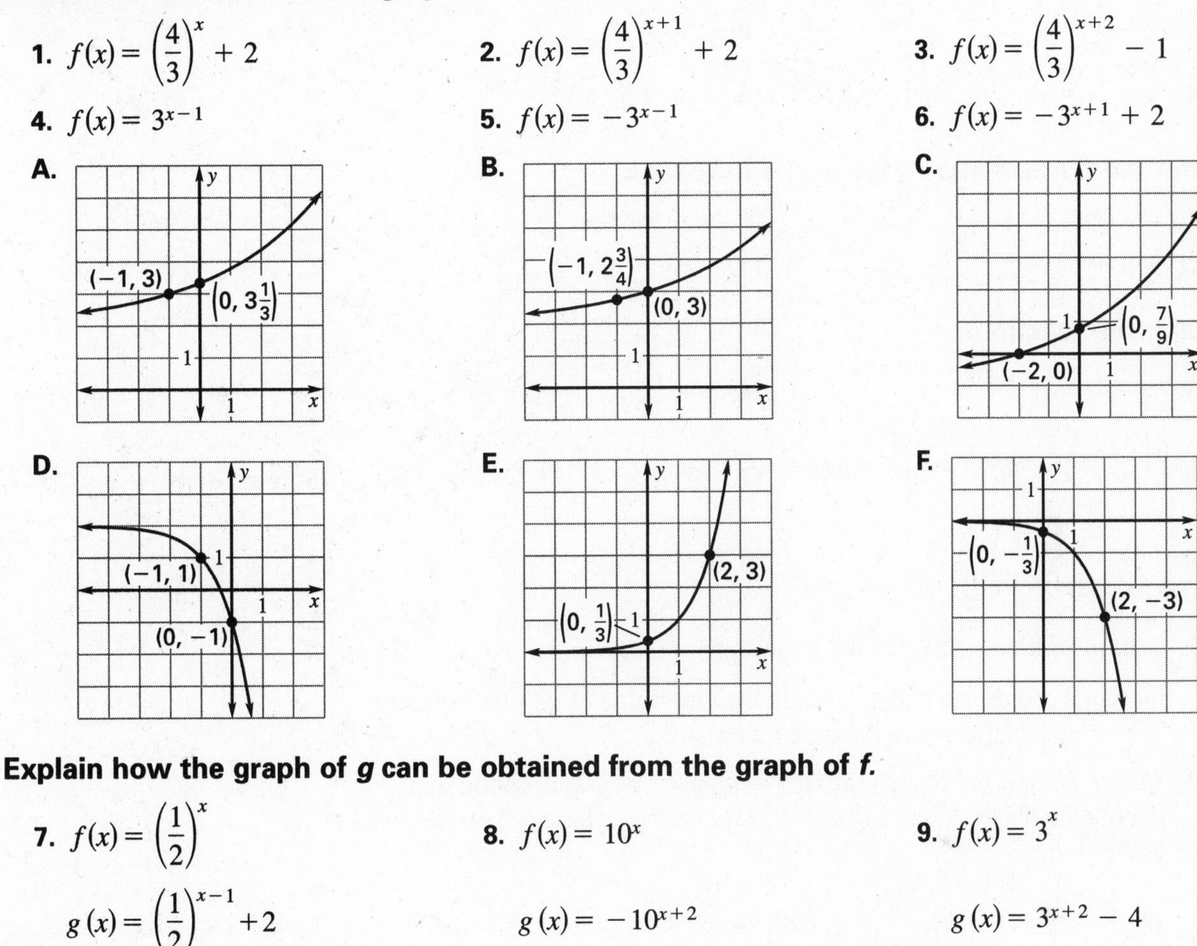

Explain how the graph of *g* can be obtained from the graph of *f*.

7. $f(x) = \left(\frac{1}{2}\right)^x$

   $g(x) = \left(\frac{1}{2}\right)^{x-1} + 2$

8. $f(x) = 10^x$

   $g(x) = -10^{x+2}$

9. $f(x) = 3^x$

   $g(x) = 3^{x+2} - 4$

Identify the *y*-intercept and the asymptote of the graph of the function.

10. $y = 3^x + 2$
11. $y = 3^{x-3}$
12. $y = 3^{x+1} - 2$

Graph the function.

13. $y = 4^{x+2}$
14. $y = 2^{x-3}$
15. $y = 3^x + 1$
16. $y = 2^x - 3$
17. $y = 3^{x+1} + 2$
18. $y = 2^{x-1} - 3$
19. $y = 2^{x+1} - 4$
20. $y = 3^{x-2} + 1$
21. $y = \left(\frac{3}{2}\right)^{x+2} - 1$

Computer Usage **In Exercises 22–24, use the following information.**

From 1991 through 1995, the number of computers C per 100 people worldwide can be modeled by $C = 25.2(1.15)^t$ where t is the number of years since 1991.

22. Identify the initial amount, the growth factor, and the annual percent increase.
23. Graph the function.
24. Estimate the number of computers per 1000 people worldwide in 2000.

LESSON 8.1

NAME ______________________ DATE ____________

Practice C

For use with pages 465–472

Identify the *y*-intercept and asymptote of the graph of the function.

1. $y = 5^x - 2$
2. $y = 5^{x-1}$
3. $y = 3(5^x) + 4$
4. $y = \frac{1}{3}(5^{x-2})$
5. $y = -3^x + 7$
6. $y = -\frac{1}{2}(5^{x-6})$

State the domain and range of the functions.

7. $y = 8^{x+1} + 3$
8. $y = 5^{x+3} - 2$
9. $y = 6^{x-1} + 4$
10. $y = 7^{x-5} - 2$
11. $y = 3(2^x) + 4$
12. $y = -2(3^{x+1}) - 3$

Graph the function.

13. $y = 4^{x-2} + 1$
14. $y = \left(\frac{3}{2}\right)^{x+3} - 2$
15. $y = 2(2^{x-1}) + 4$
16. $y = -3(2^{x+1}) + 4$
17. $y = 3\left(\frac{3^{x-2}}{2}\right) - 1$
18. $y = 3^{x-1/2} + 1$
19. $y = 2^{x+3/2} + \frac{1}{3}$
20. $y = -\frac{1}{2}(2^{x-1}) + 5$
21. $y = -3(2^{x-1/3}) - \frac{1}{2}$

22. ***Visual Thinking*** Sketch the graphs of $y = 2^x$, $y = 3^x$, and $y = \left(\frac{3}{2}\right)^x$ on the same coordinate plane. Explain how the value of a in the equation $y = a^x$ affects the graph. Assume that $a > 0$.

23. ***Visual Thinking*** Sketch the following pairs of graphs in the same coordinate plane. Assuming $a > 0$, explain the difference between $y = a^x$ and $y = a^{-x}$.

 a. $y = 2^x$; $y = 2^{-x}$
 b. $y = 3^x$; $y = 3^{-x}$
 c. $y = \left(\frac{4}{3}\right)^x$; $y = \left(\frac{4}{3}\right)^{-x}$

24. ***Account Balance*** You deposit \$1000 in an account that earns 2.5% annual interest. Find the balance after 3 years if this interest is compounded with the given frequency.

 a. monthly
 b. daily
 c. hourly

25. Use your results from Exercise 24 to determine if there is a limit to how much you can earn. If there is a limit, what is the maximum amount?

College Tuition **In Exercises 26–29, use the following information.**

In 1990, the tuition at a private college was \$15,000. During the next 9 years, tuition increased by about 7.2% each year.

26. Write a model giving the cost C of tuition at the college t years after 1990.
27. Graph the model.
28. Estimate the year when the tuition was \$20,000.
29. Estimate the tuition in 2010.

LESSON 

NAME ______________________ DATE ________

Reteaching with Practice

For use with pages 465–472

GOAL **Graph exponential growth functions and use exponential growth functions to model real-life situations**

VOCABULARY

An **exponential function** involves the expression b^x where the **base** b is a positive number other than 1. If $a > 0$ and $b > 1$, the function $y = ab^x$ is an exponential growth function.

An **asymptote** is a line that a graph approaches as you move away from the origin. In the exponential growth model $y = a(1 + r)^t$, y is the quantity after t years, a is the initial amount, r is the percent increase expressed as a decimal, and the quantity $1 + r$ is called the **growth factor.**

Compound Interest Consider an initial principal P deposited in an account that pays interest at an annual rate r (expressed as a decimal), compounded n times per year. The amount A in the account after t years can be modeled by this equation: $A = P\left(1 + \frac{r}{n}\right)^{nt}$

Lesson 8.1

EXAMPLE 1 *Graphing Exponential Functions*

Graph the function (a) $y = -2 \cdot 3^x$ and (b) $y = 2 \cdot 3^x$.

SOLUTION

Begin by plotting two points on the graph. To find these two points, evaluate the function when $x = 0$ and $x = 1$.

a. $y = -2 \cdot 3^0 = -2 \cdot 1 = -2$

$y = -2 \cdot 3^1 = -2 \cdot 3 = -6$

Plot $(0, -2)$ and $(1, -6)$. Then, from left to right, draw a curve that begins just below the x-axis, passes through the two points, and moves down to the right.

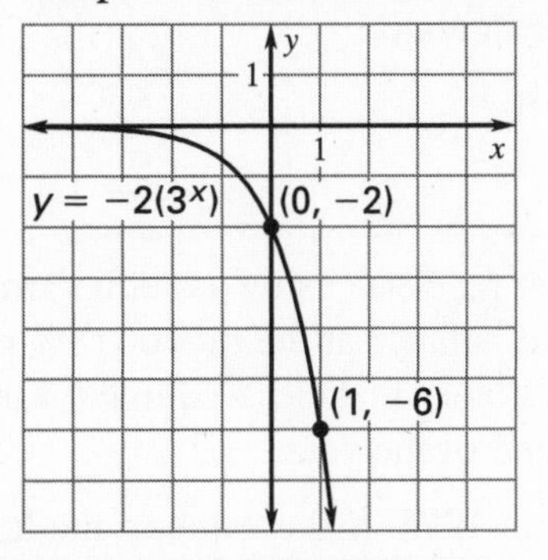

b. $y = 2 \cdot 3^0 = 2 \cdot 1 = 2$

$y = 2 \cdot 3^1 = 2 \cdot 3 = 6$

Plot $(0, 2)$ and $(1, 6)$. Then, from left to right, draw a curve that begins just above the x-axis, passes through the two points, and moves up to the right.

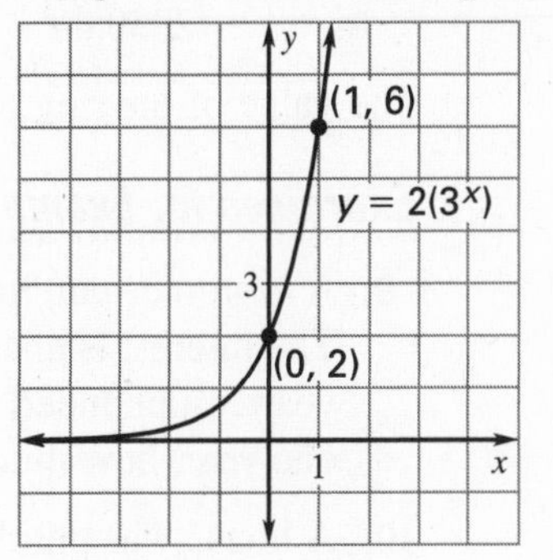

Exercises for Example 1

Graph the function.

1. $y = 2^x$

2. $y = -4^x$

3. $y = -3 \cdot 2^x$

4. $y = 4 \cdot 2^x$

Reteaching with Practice

For use with pages 465–472

EXAMPLE 2 Graphing a General Exponential Function

Graph $y = -2 \cdot 4^{x+1} + 3$. State the domain and range.

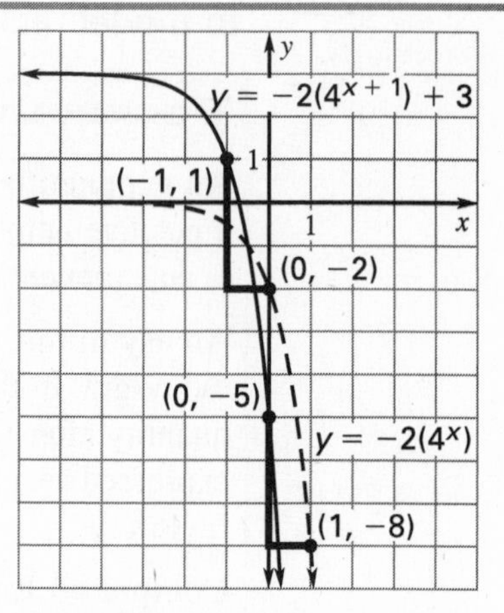

SOLUTION

Begin by lightly sketching the graph of $y = -2 \cdot 4^x$, which passes through $(0, -2)$ and $(1, -8)$. Then because $h = -1$ and $k = 3$, translate the graph 1 unit to the left and 3 units up. Notice that the graph passes through $(-1, 1)$ and $(0, -5)$. The graph's asymptote is $y = 3$. The domain is all real numbers and the range is $y < 3$.

Exercises for Example 2

Graph the function. State the domain and range.

5. $y = -3 \cdot 2^{x+4}$ **6.** $y = 5 \cdot 2^{x-1}$ **7.** $y = 3^{x-2} + 4$ **8.** $y = 4^{x+2} - 3$

EXAMPLE 3 Modeling Exponential Growth

A diamond ring was purchased twenty years ago for $500. The value of the ring increased by 8% each year. What is the value of the ring today?

SOLUTION

The initial amount is $a = 500$, the percent increase expressed in decimal form is $r = 0.08$, and the time in years is $t = 20$.

$y = a(1 + r)^t$	Write exponential growth model.
$= 500(1 + 0.08)^{20}$	Substitute $a = 500$, $r = 0.08$, and $t = 20$.
$= 500 \cdot 1.08^{20}$	Simplify.
≈ 2330.48	Use a calculator.

The value of the ring today is about $2330.48.

Exercises for Example 3

9. A customer purchases a television set for $800 using a credit card. The interest is charged on any unpaid balance at the rate of 18% per year compounded monthly. If the customer makes no payment for one year, how much is owed at the end of the year?

10. A house was purchased for $90,000 in 1995. If the value of the home increases by 5% per year, what is it worth in the year 2020?

NAME ______________________________ DATE ________

Quick Catch-Up for Absent Students

For use with pages 465–472

The items checked below were covered in class on (date missed) ____________

Lesson 8.1: Exponental Growth

____ **Goal 1:** Graph exponential growth functions (pp. 465, 466)

Material Covered:

____ Activity: Investigating Graphs of Exponential Functions

____ Student Help: Look Back

____ Example 1: Graphing Exponential Functions

____ Example 2: Graphing a General Exponential Function

Vocabulary:

exponential function, p. 465
asymptote, p. 465
base, p. 465
exponential growth function, p. 466

____ **Goal 2:** Use exponential growth functions to model real-life situations. (pp. 467, 468)

Material Covered:

____ Example 3: Modeling Exponential Growth

____ Example 4: Finding the Balance in an Account

Vocabulary:

growth factor, p. 467

____ Other (specify) ____

____

Homework and Additional Learning Support

____ Textbook (specify) pp. 469–472

____

____ Internet: Extra Examples at www.mcdougallittell.com

____ *Reteaching with Practice* worksheet (specify exercises)________________

____ *Personal Student Tutor* for Lesson 8.1

NAME ______________________________ DATE __________

Real-Life Application: When Will I Ever Use This?

For use with pages 465–472

Cellular Telephones

The first call on a commercial cellular system in the United States was made in Chicago, Illinois, in 1983. Since then, the cellular telephone industry has been one of the fastest growing industries in the United States. Today, there are more than seven hundred times as many cellular telephone subscribers than there were in 1984, and the number keeps growing.

A cellular communication system consists of three main parts: the cellular telephone itself, a network of radio antennas called base stations, and a central switching office. Service areas are divided into geographic zones called cells, where a base station is usually centrally located. When someone makes a call on a cellular phone, radio waves carry the message to the base station, the base station sends them to the central switching office, and from there, the message is sent to the local or long-distance telephone company, where it is sent to its destination. Base stations are set up over a geographical area based on the network. As a cellular phone user moves across a service area, their call is transferred from base station to base station.

Cellular telephones have become a convenient and useful way of communicating, whether you are talking to a friend or doing business.

In Exercises 1–6, use the following information.

From 1990 to 1997, the number of cellular telephone subscribers S (in thousands) in the United States can be modeled by

$$S = 5535.33(1.413)^t$$

where t is the number of years since 1990.

1. Identify the growth factor and annual percent increase.
2. Sketch a graph of the model.
3. In what year was the number of cellular telephone subscribers about 31 million?
4. According to the model, in what year will the number of cellular telephone subscribers exceed 90 million?
5. Estimate the number of subscribers in 2002, 2005, and 2010.
6. Do you think this model can be used to predict the future number of cellular telephone subscribers? Explain.

LESSON 8.1

NAME ______________________________ DATE ______________

Challenge: Skills and Applications

For use with pages 465–472

In Exercises 1–4, write each function in the form $y = ab^x$.

1. $y = 4^{x+1}$

2. $y = 5 \cdot \sqrt{9^{x-1}}$

3. $y = \sqrt{5 \cdot 2^{3x}}$

4. $y = \dfrac{\sqrt{2}}{3 \cdot 5^{2x}}$

5. A *mortgage* is a loan that is paid off in equal monthly installments M, which include interest as well as repayment of the principal. The amount p_k of the kth installment that goes toward repayment of principal is given by the formula

$$p_k = \frac{M}{(1 + r)^{n-k+1}}$$

where the mortgage is fully repaid in n months and r is the monthly interest rate.

a. Suppose $r = 0.7\%$, $M = \$1050$, and $n = 300$. Find p_1 and p_{300}.

b. Is the formula a model of exponential growth?

6. a. Let $f(x) = 2^x$. Find the finite (first) differences between the values of f for $x = 0$, 1, 2, 3, 4, and 5. Write a formula that gives the difference between $f(x + 1)$ and $f(x)$ in terms of x.

b. Repeat part (a) but use the function $y = 3^x$.

c. Repeat part (a) for the function $y = 5^x$.

d. Make a conjecture for a general formula that gives the difference between $f(x + 1)$ and $f(x)$ for the function $y = a^x$, for $a > 1$.

7. The *average growth rate* of an exponential function $f(x)$ on an interval $a \le x \le b$ is

$$\frac{f(b) - f(a)}{b - a}.$$

The *instantaneous growth rate at a* is the number that this fraction approaches (if any) as $b \to a$. Let $f(x) = 3^x$, and let $b = a + h$. Write the instantaneous growth rate at a as the product of 3^a and a function of h (that does not involve a). Use a calculator to estimate the number this second factor approaches as $h \to 0$.

LESSON 8.2

TEACHER'S NAME ______________________ CLASS ________ ROOM ______ DATE ______

Lesson Plan

1-day lesson (See *Pacing the Chapter,* TE pages 462C–462D) **For use with pages 473–479**

GOALS

1. **Graph exponential decay functions.**
2. **Use exponential decay functions to model real-life situations.**

State/Local Objectives ____

____

✓ Check the items you wish to use for this lesson.

STARTING OPTIONS

____ Homework Check: TE page 469; Answer Transparencies
____ Warm-Up or Daily Homework Quiz: TE pages 474 and 472, CRB page 24, or Transparencies

TEACHING OPTIONS

____ Concept Activity: SE page 473; CRB page 25 (Activity Support Master)
____ Lesson Opener (Visual Approach): CRB page 26 or Transparencies
____ Graphing Calculator Activity with Keystrokes: CRB pages 27–28
____ Examples 1–4: SE pages 474–476
____ Extra Examples: TE pages 475–476 or Transparencies; Internet
____ Closure Question: TE page 476
____ Guided Practice Exercises: SE page 477

APPLY/HOMEWORK

Homework Assignment

____ Basic 12–40 even, 44, 50–52, 57–67 odd
____ Average 12–40 even, 44, 47–52, 57–67 odd
____ Advanced 12–40 even, 44, 47–58, 59–67 odd

Reteaching the Lesson

____ Practice Masters: CRB pages 29–31 (Level A, Level B, Level C)
____ Reteaching with Practice: CRB pages 32–33 or Practice Workbook with Examples
____ Personal Student Tutor

Extending the Lesson

____ Applications (Interdisciplinary): CRB page 35
____ Challenge: SE page 479; CRB page 36 or Internet

ASSESSMENT OPTIONS

____ Checkpoint Exercises: TE pages 475–476 or Transparencies
____ Daily Homework Quiz (8.2): TE page 479, CRB page 39, or Transparencies
____ Standardized Test Practice: SE page 479; TE page 479; STP Workbook; Transparencies

Notes ____

____

____

TEACHER'S NAME ______________________________ CLASS ________ ROOM ________ DATE ________

Lesson Plan for Block Scheduling

Half-day lesson (See *Pacing the Chapter,* TE pages 462C–462D) **For use with pages 473–479**

1. Graph exponential decay functions.
2. Use exponential decay functions to model real-life situations.

State/Local Objectives ____

____

____

CHAPTER PACING GUIDE	
Day	**Lesson**
1	8.1 (all); **8.2(all)**
2	8.3 (all)
3	8.4 (all)
4	8.5 (all)
5	8.6 (all)
6	8.7 (all); 8.8(all)
7	Review/Assess Ch. 8

✓ Check the items you wish to use for this lesson.

STARTING OPTIONS

____ Homework Check: TE page 469; Answer Transparencies
____ Warm-Up or Daily Homework Quiz: TE pages 474 and 472, CRB page 24, or Transparencies

TEACHING OPTIONS

____ Concept Activity: SE page 473; CRB page 25 (Activity Support Master)
____ Lesson Opener (Visual Approach): CRB page 26 or Transparencies
____ Graphing Calculator Activity with Keystrokes: CRB pages 27–28
____ Examples 1–4: SE pages 474–476
____ Extra Examples: TE pages 475–476 or Transparencies; Internet
____ Closure Question: TE page 476
____ Guided Practice Exercises: SE page 477

APPLY/HOMEWORK

Homework Assignment (See also the assignment for Lesson 8.1.)

____ Block Schedule: 12–40 even, 44, 47–52, 57–67 odd

Reteaching the Lesson

____ Practice Masters: CRB pages 29–31 (Level A, Level B, Level C)
____ Reteaching with Practice: CRB pages 32–33 or Practice Workbook with Examples
____ Personal Student Tutor

Extending the Lesson

____ Applications (Interdisciplinary): CRB page 35
____ Challenge: SE page 479; CRB page 36 or Internet

ASSESSMENT OPTIONS

____ Checkpoint Exercises: TE pages 475–476 or Transparencies
____ Daily Homework Quiz (8.2): TE page 479, CRB page 39, or Transparencies
____ Standardized Test Practice: SE page 479; TE page 479; STP Workbook; Transparencies

Notes ____

____

____

LESSON 8.2

Available as a transparency

NAME ______________________ DATE ________

WARM-UP EXERCISES

For use before Lesson 8.2, pages 473–479

Identify the value of *b* in each exponential function, $f(x) = ab^x$.

1. $f(x) = 3 \cdot \left(\frac{1}{2}\right)^x$
2. $f(x) = 3.5^x$
3. $f(x) = 5 \cdot (-2)^{x+1}$

State the domain and range of each function.

4. $y = 5 \cdot 3^x$
5. $y = -\frac{1}{4} \cdot 2^x$

Lesson 8.2

DAILY HOMEWORK QUIZ

For use after Lesson 8.1, pages 465–472

1. Identify the y-intercept and the asymptotes of the graph of $y = 0.5 \cdot 3^{x+2}$.
2. Graph $y = -1.5 \cdot 2^x$.
3. Graph $y = 2^{x+1} - 3$. State the domain and range.

NAME ______________________ DATE __________

Activity Support Master

For use with page 473

Fold number	1	2	3	4	5
Number of regions	1	2			
Fractional area of each region	1	$\frac{1}{2}$			

First scatter plot: ***(fold number, number of regions)***

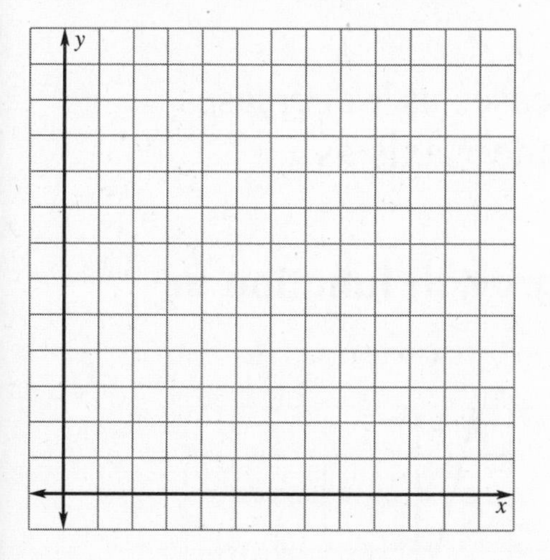

Second scatter plot: ***(fold number, fractional area of each region)***

LESSON **8.2**

NAME ______________________ DATE__________

Available as a transparency

Visual Approach Lesson Opener

For use with pages 474–479

Two important kinds of functions are *exponential growth functions* and *exponential decay functions.*

Exponential growth: $y = ab^x$, where $a > 0$ and $b > 1$
For these functions, y increases as x increases and the graph approaches the x-axis as x decreases.

Exponential decay: $y = ab^x$, where $a > 0$ and $0 < b < 1$
For these functions, y increases as x decreases and the graph approaches the x-axis as x increases.

Classify each function as an exponential growth function or an exponential decay function.

1.

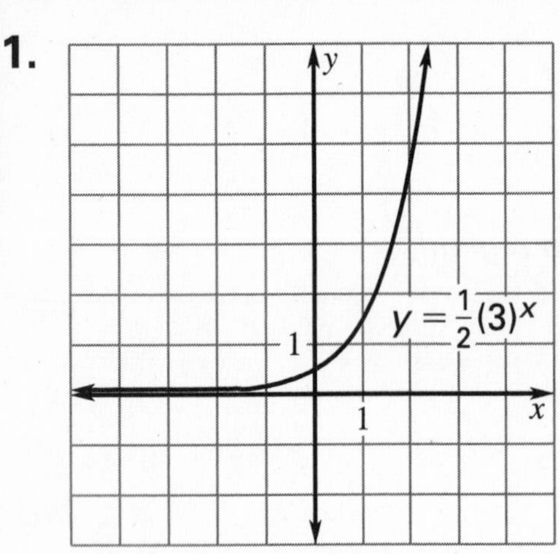

2.

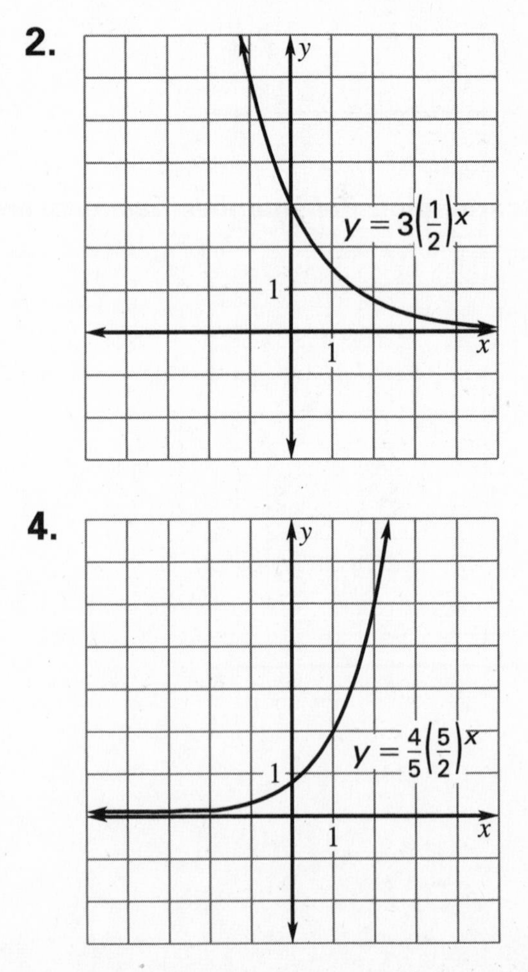

3.

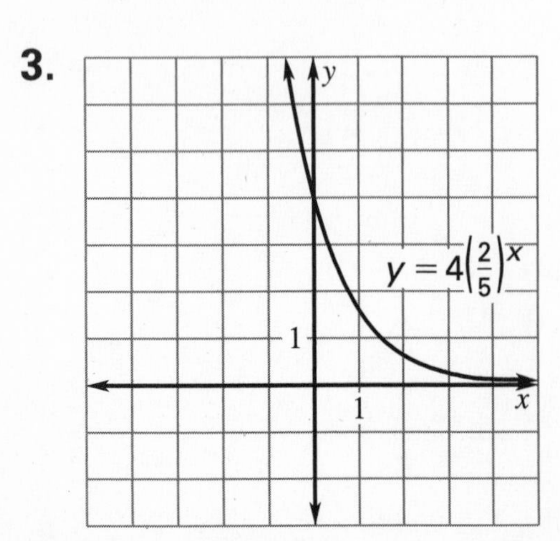

4.

NAME ______________________ DATE ________

Graphing Calculator Activity

For use with pages 474–479

GOAL **To compare the graph of the exponential decay function to that of the exponential growth function**

Recall that *exponential growth functions* are of the form $f(x) = ab^x$, where $a > 0$ and $b > 1$. *Exponential decay functions* are of the form $f(x) = ab^x$, where $a > 0$ and $0 < b < 1$.

Activity

1. Use a graphing calculator to graph the following.

   $y = 3^x$, $y = 2 \cdot 3^x$, $y = 2 \cdot 3^{x+2}$

2. Use a graphing calculator to graph the following.

   $y = \left(\frac{1}{3}\right)^x$, $y = 2\left(\frac{1}{3}\right)^x$, $y = 2\left(\frac{1}{3}\right)^{x+2}$

3. Describe how the graphs of the exponential growth functions in Step 1 differ from those of the exponential decay functions in Step 2.

Exercises

1. Will each of the following graphs rise or fall from left to right?

$y = 5^x$, $y = \left(\frac{1}{5}\right)^x$, $y = 0.5^x$, $y = 7^x$, $y = 3 \cdot 5^x$

2. Check your answers to Exercise 1 with a graphing calculator.

3. Explain how you know the graph of an exponential function rises or falls from left to right without graphing the function.

4. Without graphing, determine whether the function rises or falls from left to right.

(a) $y = \left(\frac{1}{4}\right)^x$ (b) $y = 2\left(\frac{1}{4}\right)^x$ (c) $y = 2 \cdot 5^{x+2} + 1$

(d) $y = \left(\frac{1}{7}\right)^x - 3$ (e) $y = 3^x - 2$ (f) $y = 2 \cdot 7^x - 1$

5. Check your answers to Exercise 4 with a graphing calculator.

NAME _______________________________ DATE ___________

Graphing Calculator Activity

For use with pages 474–479

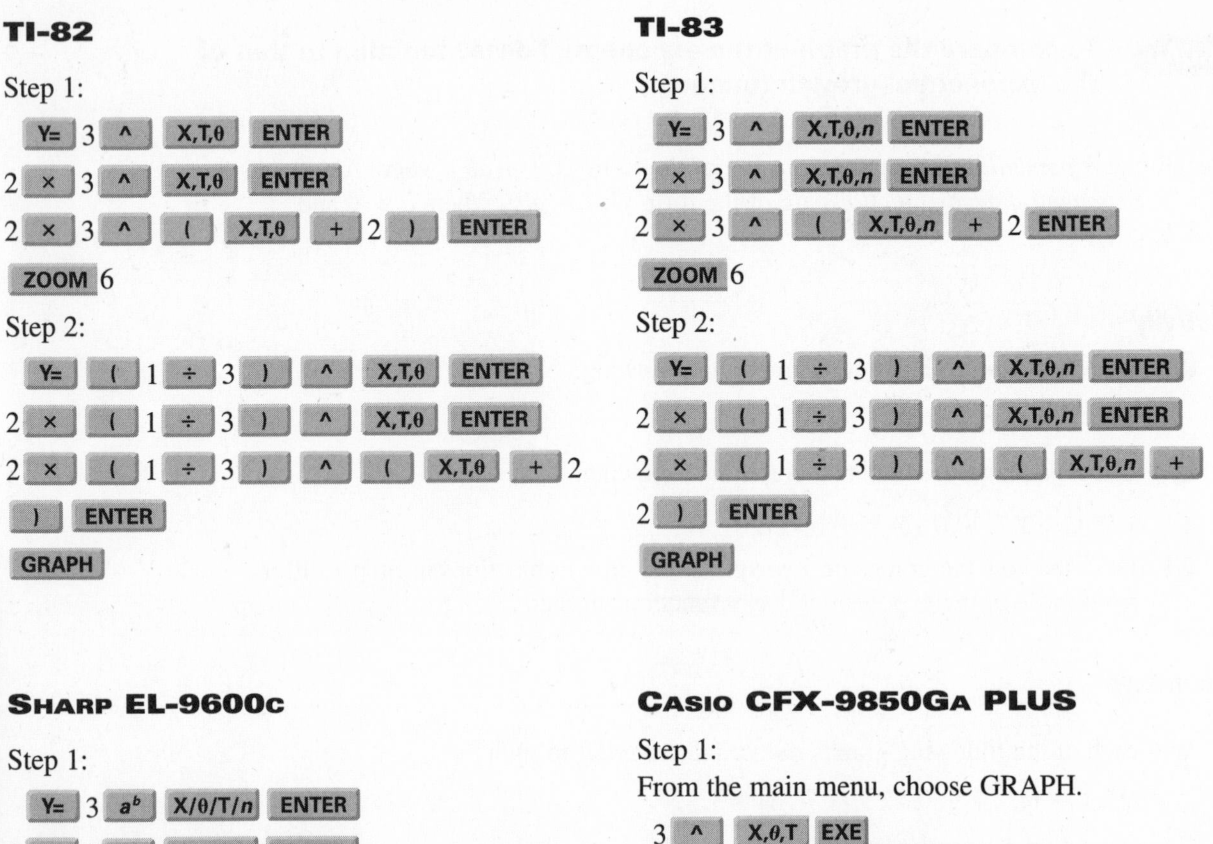

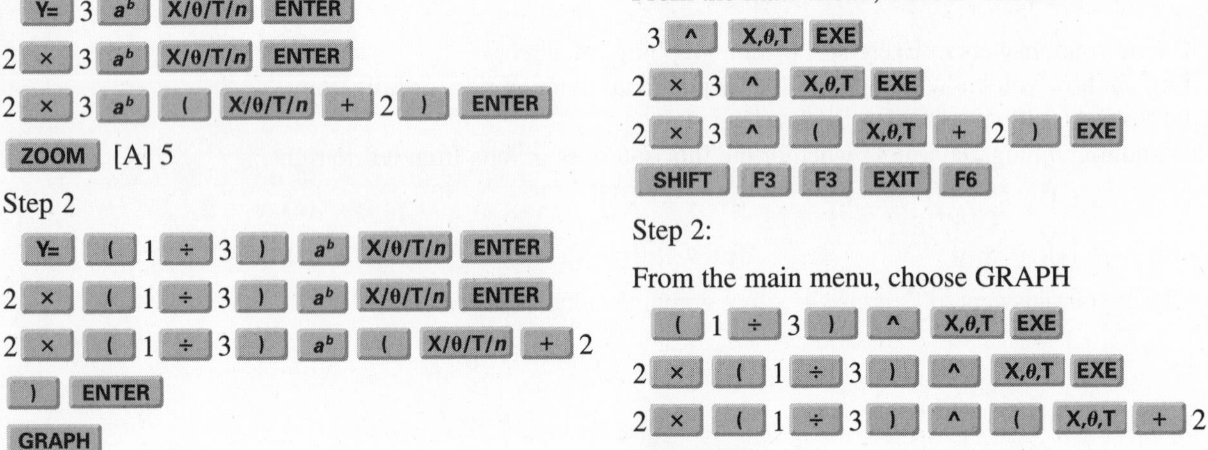

) EXE

F6

LESSON **8.2**

NAME ______________________ DATE __________

Practice A

For use with pages 474–479

Tell whether the function represents *exponential growth* or *exponential decay*.

1. $f(x) = \left(\frac{2}{3}\right)^x$
2. $f(x) = \left(\frac{5}{4}\right)^x$
3. $f(x) = 6^x$
4. $f(x) = (0.7)^x$
5. $f(x) = \left(\frac{1}{3}\right)^x$
6. $f(x) = \frac{1}{2}(3^x)$

Match the function with its graph.

7. $y = \left(\frac{1}{3}\right)^x$
8. $y = -\left(\frac{1}{3}\right)^x$
9. $y = 2\left(\frac{1}{3}\right)^x$

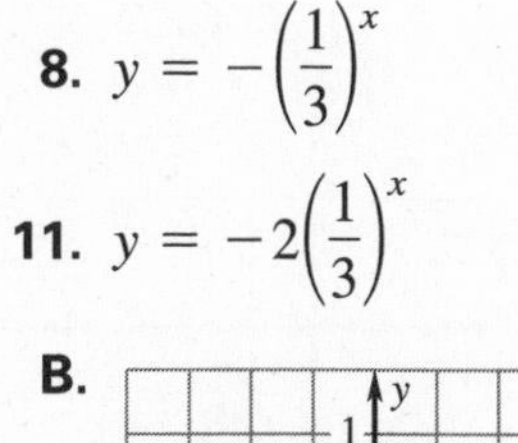

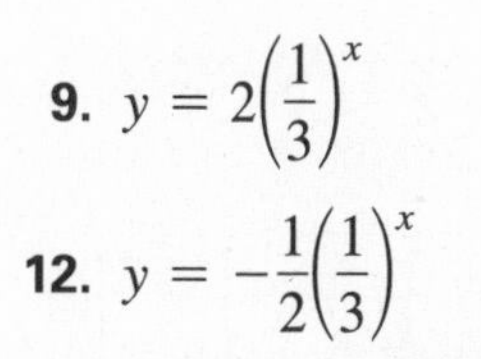

10. $y = \frac{1}{2}\left(\frac{1}{3}\right)^x$
11. $y = -2\left(\frac{1}{3}\right)^x$
12. $y = -\frac{1}{2}\left(\frac{1}{3}\right)^x$

A.

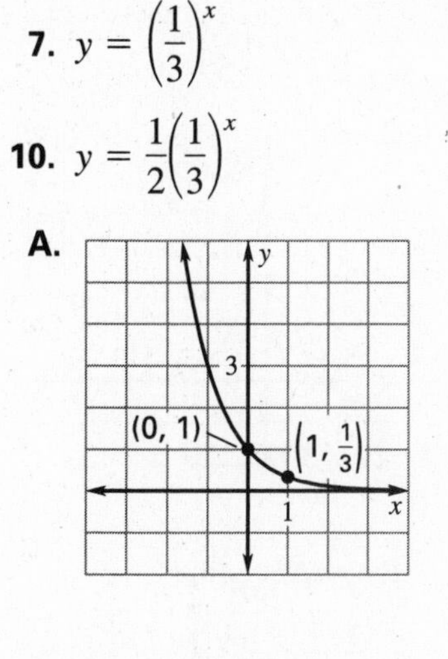

B.

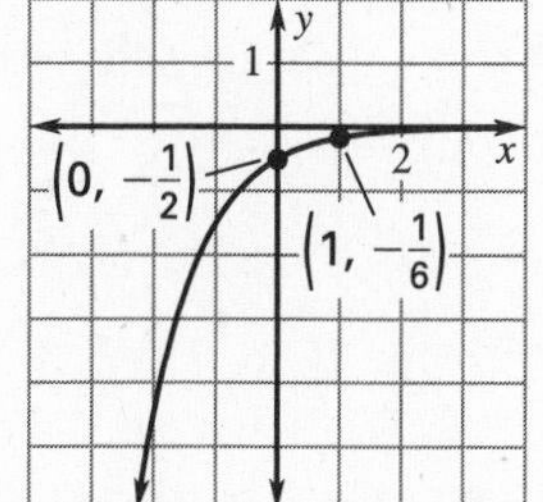

C.

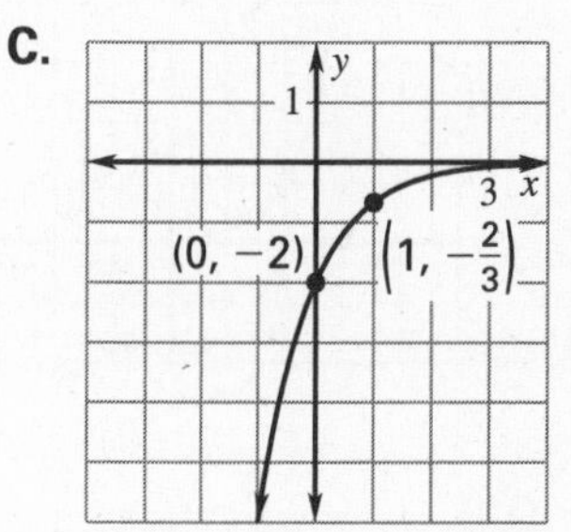

D.

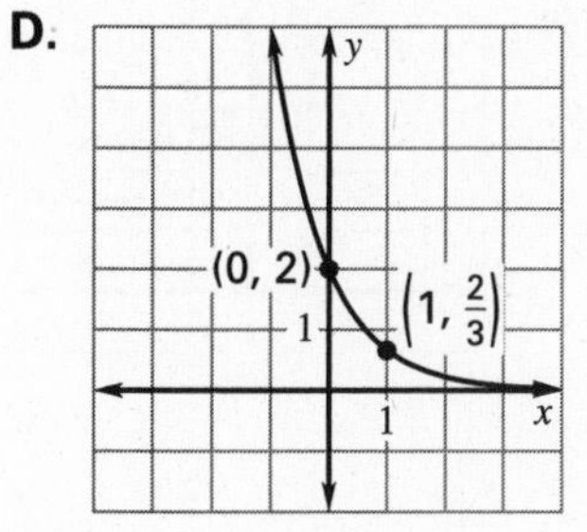

E.

F.

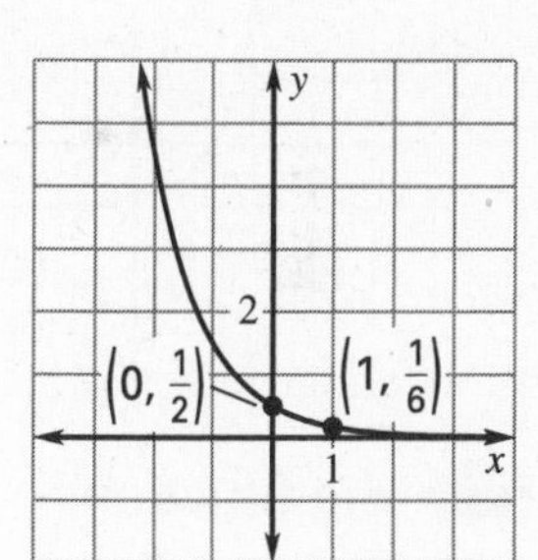

Identify the *y*-intercept and asymptote of the graph of the function.

13. $y = \left(\frac{2}{3}\right)^x$
14. $y = (0.3)^x$
15. $y = 2\left(\frac{1}{3}\right)^x$
16. $y = \frac{1}{4}\left(\frac{8}{9}\right)^x$
17. $y = -5\left(\frac{1}{2}\right)^x$
18. $y = -\frac{2}{3}\left(\frac{1}{5}\right)^x$

19. ***Radioactive Decay*** Ten grams of Carbon 14 is stored in a container. The amount C (in grams) of Carbon 14 present after t years can be modeled by $C = 10(0.99987)^t$. How much Carbon 14 is present after 1000 years?

Lesson 8.2

NAME ______________________________________ DATE __________

Practice B

For use with pages 474–479

Tell whether the function represents *exponential growth* or *exponential decay*.

1. $f(x) = \frac{1}{2}\left(\frac{5}{7}\right)^x$

2. $f(x) = \frac{1}{3}\left(\frac{7}{5}\right)^x$

3. $f(x) = 3(4)^{-x}$

Match the function with its graph.

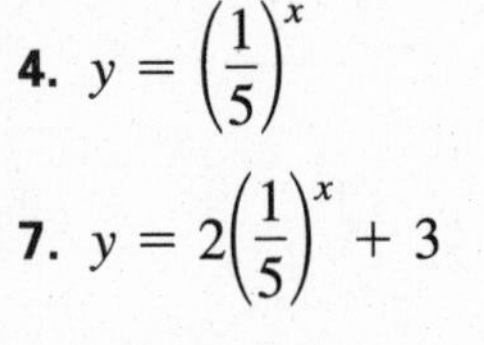

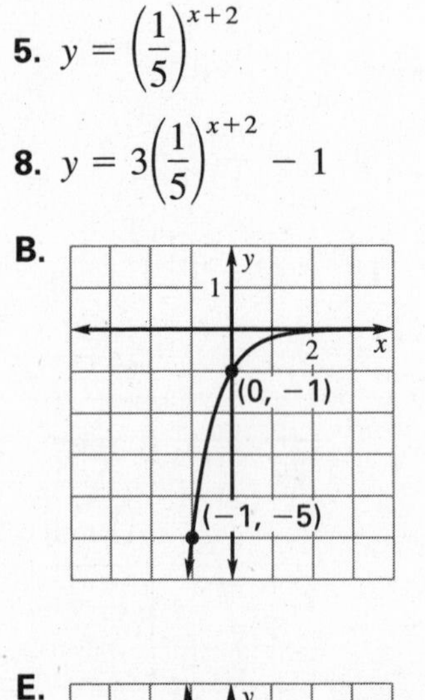

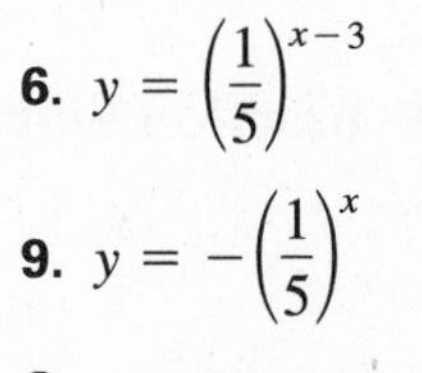

4. $y = \left(\frac{1}{5}\right)^x$

5. $y = \left(\frac{1}{5}\right)^{x+2}$

6. $y = \left(\frac{1}{5}\right)^{x-3}$

7. $y = 2\left(\frac{1}{5}\right)^x + 3$

8. $y = 3\left(\frac{1}{5}\right)^{x+2} - 1$

9. $y = -\left(\frac{1}{5}\right)^x$

A.

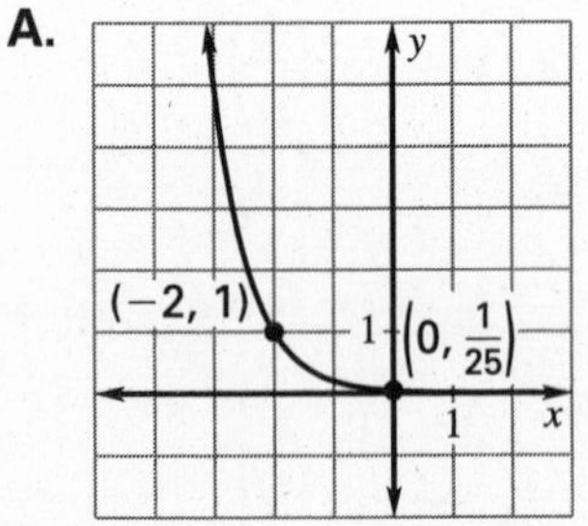

B.

C.

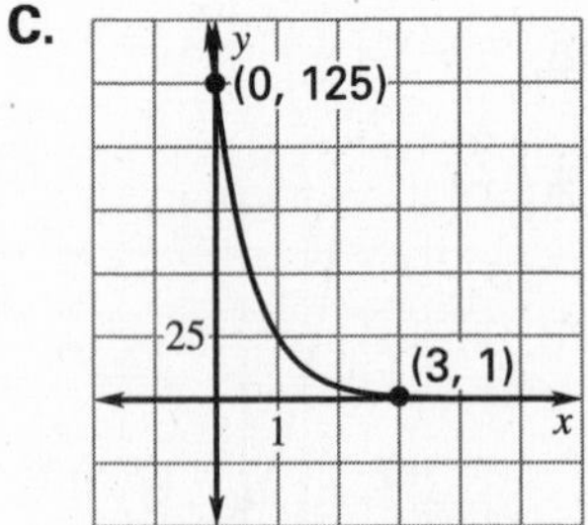

D.

E.

F.

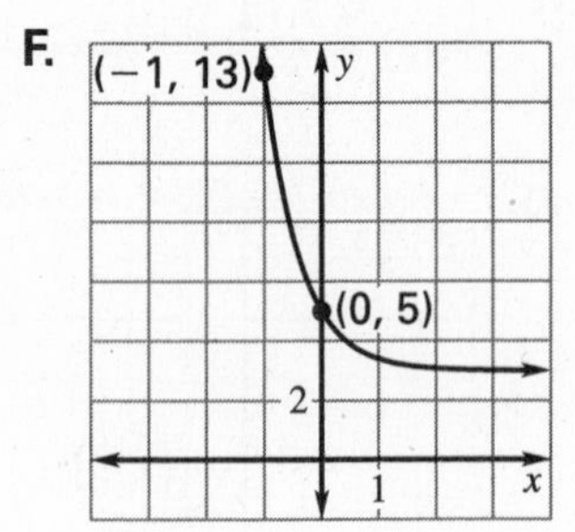

Graph the function.

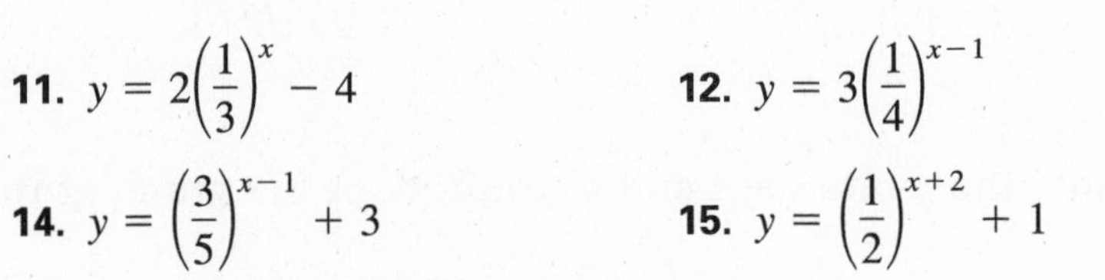

10. $y = 2\left(\frac{1}{2}\right)^x + 3$

11. $y = 2\left(\frac{1}{3}\right)^x - 4$

12. $y = 3\left(\frac{1}{4}\right)^{x-1}$

13. $y = 2\left(\frac{1}{5}\right)^{x+3}$

14. $y = \left(\frac{3}{5}\right)^{x-1} + 3$

15. $y = \left(\frac{1}{2}\right)^{x+2} + 1$

Value of the Dollar **In Exercises 16–18, use the following information.**

From 1990 through 1998, the value of the dollar has been shrinking. That is, you cannot buy as much with a dollar today as you could in 1990. The shrinking value can be modeled by $V = 1.24(0.973)^t$, where t is the number of years since 1990.

16. How much was a 1998 dollar worth in 1993?

17. Graph the model.

18. Estimate the year in which the 1998 dollar was worth \$1.07.

NAME ______________________ DATE __________

Practice C

For use with pages 474–479

Tell whether the function represents *exponential growth* or *exponential decay*.

1. $f(x) = \left(\frac{2}{3}\right)^x$
2. $f(x) = \left(\frac{3}{2}\right)^x$
3. $f(x) = \left(\frac{3}{2}\right)^{-x}$
4. $f(x) = \left(\frac{2}{3}\right)^{-x}$
5. $f(x) = -\left(\frac{3}{2}\right)^{-x}$
6. $f(x) = -\left(\frac{2}{3}\right)^{-x}$

Identify the *y*-intercept and asymptote of the graph of the function.

7. $y = \left(\frac{1}{2}\right)^x + 3$
8. $y = \left(\frac{2}{3}\right)^{x+3}$
9. $y = \frac{1}{4}\left(\frac{3}{4}\right)^{x+1}$

State the domain and range of the function.

10. $y = \left(\frac{1}{2}\right)^{x+1} - 3$
11. $y = \left(\frac{1}{3}\right)^{x-2} + 4$
12. $y = \left(\frac{2}{5}\right)^{x+4} + 1$
13. $y = \left(\frac{3}{5}\right)^{x-3} - 2$
14. $y = 3^{-x} + 7$
15. $y = -2(3^{-x}) - 4$

Graph the function.

16. $y = \left(\frac{1}{2}\right)^{x-1} + 3$
17. $y = \left(\frac{1}{3}\right)^{x+1} - 2$
18. $y = 2\left(\frac{1}{2}\right)^{x-3} + 1$
19. $y = -3\left(\frac{2}{3}\right)^{x+1} + 2$
20. $y = 2\left(\frac{3}{4}\right)^{x-1} - 2$
21. $y = \left(\frac{1}{2}\right)^{x-1/3} + 2$
22. $y = \left(\frac{1}{2}\right)^{x+3/2} - \frac{1}{4}$
23. $y = -2\left(\frac{1}{3}\right)^{x+1} + 3$
24. $y = -3\left(\frac{2}{3}\right)^{x-1/2} - \frac{4}{3}$

Equipment Depreciation **In Exercises 25–28, use the following information.**

A tool and die business purchases a piece of equipment for $250,000. The value of the equipment depreciates at a rate of 12% each year.

25. Write an exponential decay model for the value of the equipment.
26. What is the value of the equipment after 5 years?
27. Graph the model.
28. Use the model to estimate when the equipment will have a value of $70,000.

Stereo System **In Exercises 29 and 30, use the following information.**

You purchase a stereo system for $830. After a 3 month trial period, the value of the stereo system decreases 13% each year.

29. Write an exponential decay model for the value of the stereo system in terms of the number of years since the purchase.
30. What was the value of the system after 1 year?

NAME ____________________ DATE __________

Reteaching with Practice

For use with pages 474–479

GOAL **Graph exponential decay functions and use exponential decay functions to model real-life situations**

VOCABULARY

An **exponential decay function** has the form $f(x) = ab^x$, where $a > 0$ and $0 < b < 1$.

An exponential decay model has the form $y = a(1 - r)^t$, where y is the quantity after t years, a is the initial amount, r is the percent decrease expressed as a decimal, and the quantity $1 - r$ is called the **decay factor.**

EXAMPLE 1 *Recognizing Exponential Growth and Decay*

State whether $f(x)$ is an exponential growth or exponential decay function.

a. $f(x) = -4\left(\frac{1}{3}\right)^x$ **b.** $f(x) = 5\left(\frac{3}{4}\right)^{-x}$ **c.** $f(x) = 2(0.15)^x$

SOLUTION

a. Because $b = \frac{1}{3}$, and $0 < b < 1$, f is an exponential decay function.

b. Rewrite the function without negative exponents as $f(x) = 5 \cdot \left(\frac{4}{3}\right)^x$. Because $b = \frac{4}{3}$, and $b > 1$, f is an exponential growth function.

c. Because $b = 0.15$, and $0 < b < 1$, f is an exponential decay function.

Exercises for Example 1

State whether the function represents *exponential growth* or *exponential decay.*

1. $f(x) = -3 \cdot 4^x$ **2.** $f(x) = 2 \cdot (0.75)^x$ **3.** $f(x) = 4\left(\frac{1}{3}\right)^x$

4. $f(x) = 4\left(\frac{6}{5}\right)^x$ **5.** $f(x) = 3\left(\frac{1}{4}\right)^{-x}$ **6.** $f(x) = -7\left(\frac{5}{2}\right)^{-x}$

EXAMPLE 2 *Graphing Exponential Functions*

Graph the function (a) $y = -2\left(\frac{1}{3}\right)^x$ and (b) $y = 3\left(\frac{2}{3}\right)^x$.

NAME ______________________ DATE ____________

Reteaching with Practice

For use with pages 474–479

SOLUTION

Begin by plotting two points on the graph. To find these two points, evaluate the function when $x = 0$ and $x = 1$.

a. $y = -2\left(\frac{1}{3}\right)^0 = -2$

$y = -2\left(\frac{1}{3}\right)^1 = -\frac{2}{3}$

Plot $(0, -2)$ and $\left(1, -\frac{2}{3}\right)$. Then, from *right* to *left,* draw a curve that begins just below the x-axis, passes through the two points, and moves down to the left.

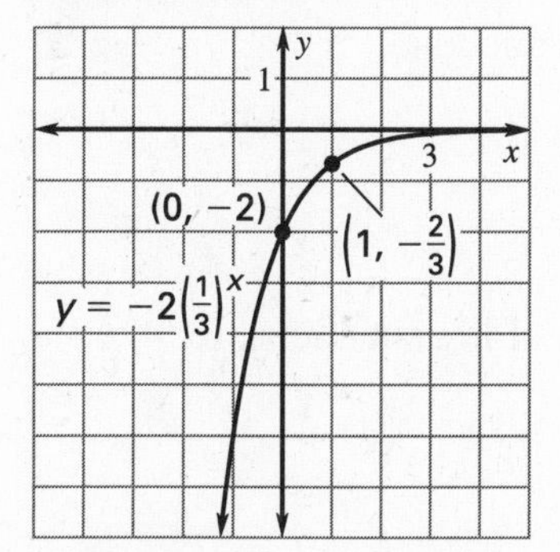

b. 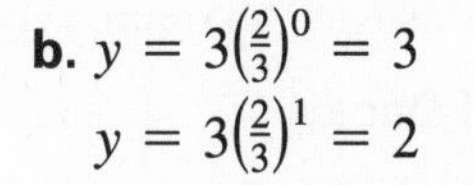
$y = 3\left(\frac{2}{3}\right)^0 = 3$

$y = 3\left(\frac{2}{3}\right)^1 = 2$

Plot $(0, 3)$ and $(1, 2)$. Then, from *right* to *left* draw a curve that begins just above the x-axis, passes through the two points, and moves up to the left.

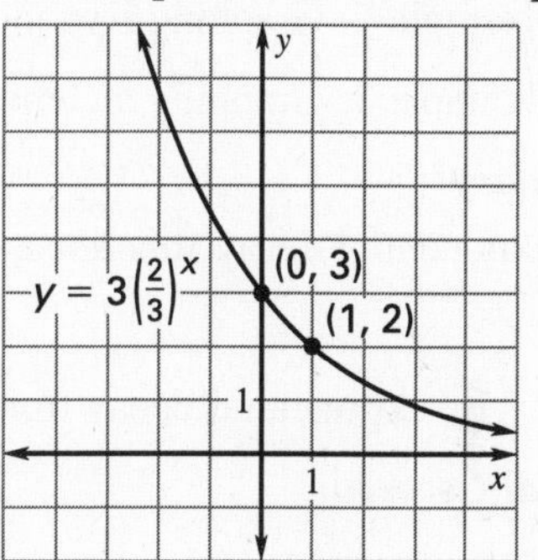

Exercises for Example 2

Graph the function.

7. $y = 2\left(\frac{1}{4}\right)^x$ **8.** $y = -3\left(\frac{1}{2}\right)^x$ **9.** $y = 4\left(\frac{3}{4}\right)^x$ **10.** $y = -5\left(\frac{2}{3}\right)^x$

EXAMPLE 3 *Graphing a General Exponential Function*

Graph $y = 2\left(\frac{1}{3}\right)^{x-4} - 5$. State the domain and range.

SOLUTION

Begin by lightly sketching the graph of $y = 2\left(\frac{1}{3}\right)^x$, which passes through $(0, 2)$ and $\left(1, \frac{2}{3}\right)$. Then, because $h = 4$ and $k = -5$, translate the graph 4 units to the right and 5 units down. Notice that the graph passes through $(4, -3)$ and $\left(5, -4\frac{1}{3}\right)$. The graph's asymptote is the line $y = -5$. The domain is all real numbers and the range is $y > -5$.

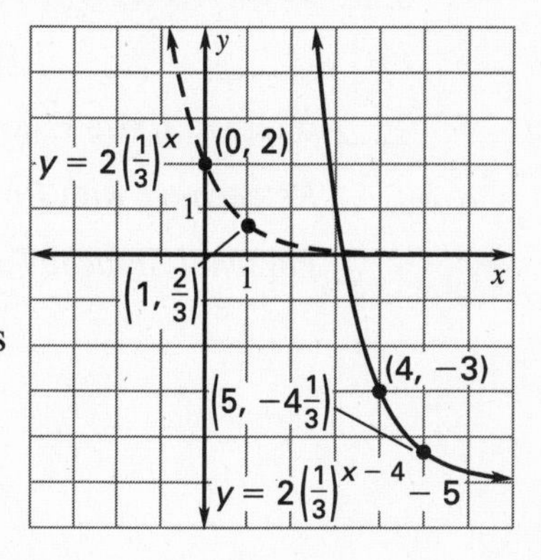

Exercises for Example 3

Graph the function. State the domain and range.

11. $y = 2\left(\frac{1}{2}\right)^{x+3}$ **12.** $y = -3\left(\frac{2}{3}\right)^{x-4}$

13. $y = -\left(\frac{1}{4}\right)^x + 2$ **14.** $y = 4\left(\frac{1}{2}\right)^{x+4} - 3$

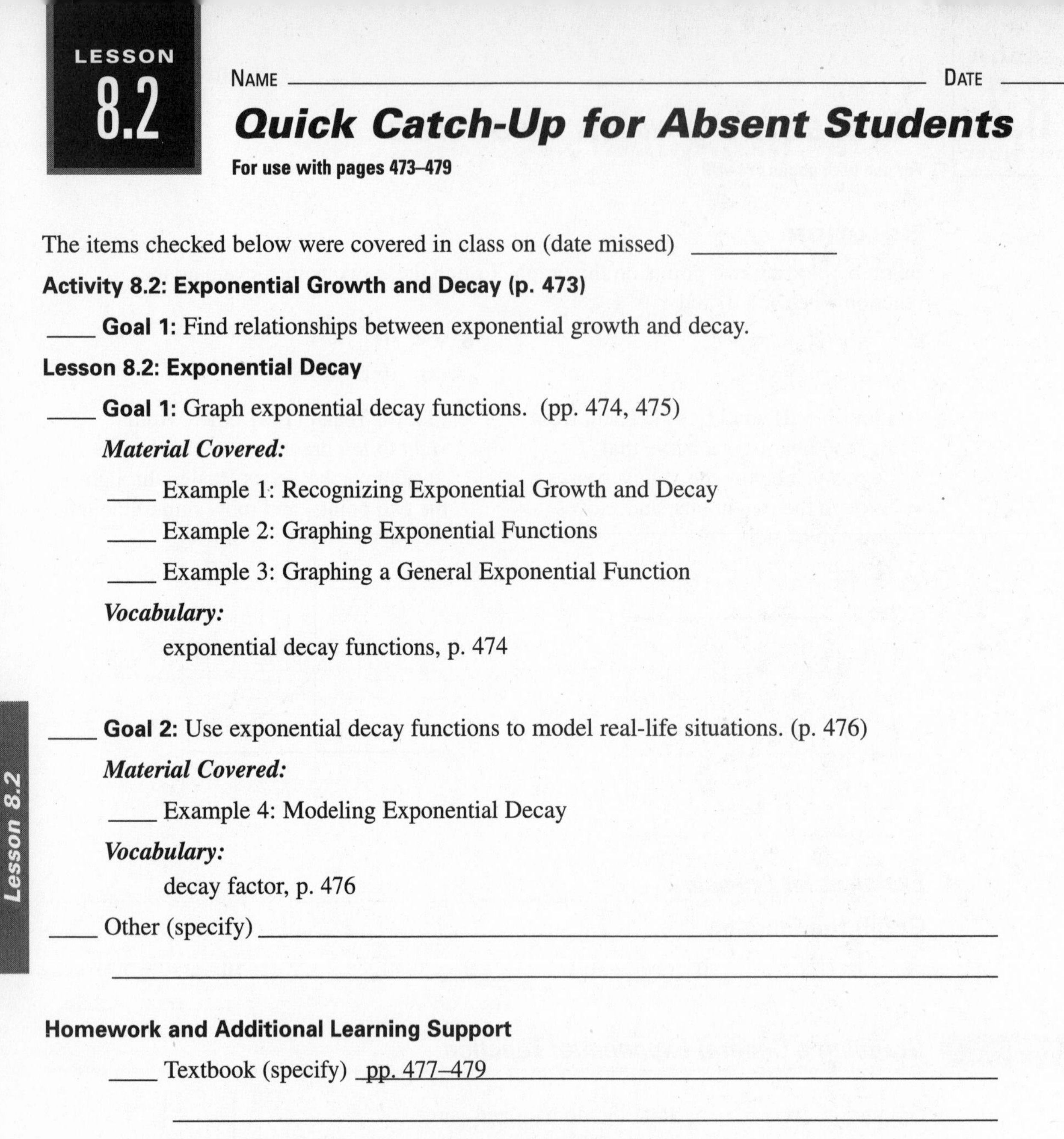

LESSON 8.2

NAME ______________________________ DATE __________

Quick Catch-Up for Absent Students

For use with pages 473–479

The items checked below were covered in class on (date missed) __________

Activity 8.2: Exponential Growth and Decay (p. 473)

____ **Goal 1:** Find relationships between exponential growth and decay.

Lesson 8.2: Exponential Decay

____ **Goal 1:** Graph exponential decay functions. (pp. 474, 475)

Material Covered:

____ Example 1: Recognizing Exponential Growth and Decay

____ Example 2: Graphing Exponential Functions

____ Example 3: Graphing a General Exponential Function

Vocabulary:

exponential decay functions, p. 474

____ **Goal 2:** Use exponential decay functions to model real-life situations. (p. 476)

Material Covered:

____ Example 4: Modeling Exponential Decay

Vocabulary:

decay factor, p. 476

____ Other (specify) ______________________________

____

Homework and Additional Learning Support

____ Textbook (specify) pp. 477–479

____

____ Internet: Extra Examples at www.mcdougallittell.com

____ *Reteaching with Practice* worksheet (specify exercises) __________

____ *Personal Student Tutor* for Lesson 8.2

NAME ______________________ DATE __________

Interdisciplinary Application

For use with pages 474–479

Tuberculosis

HEALTH Tuberculosis, commonly known as "TB", was once a common cause of death in the world. Today, improved methods of diagnosis and treatment have greatly reduced the number of people who contract the disease and the number of people who die as a result of it.

The bacteria that causes tuberculosis, tubercle bacilli, was discovered by German physician Robert Koch in 1882. The bacteria affect the body by settling into areas of a person's respiratory system. A primary infection occurs when the bacteria passes beyond a person's airways and infects the lungs. The bacteria then can multiply and infect the surrounding tissue. If the bacteria enter the bloodstream, it can spread to other parts of the body, such as the kidneys, skin, and even the brain.

In the late 1800s, one of the most common forms of treatment for tuberculosis was a sanitarium, which was a health resort where patients would go to rest and try to regain their health. Since these patients were isolated from the public, sanitariums helped to prevent the spreading of the disease as well. The downside to this form of treatment was that patients would sometimes have to spend months or even years in the sanitarium before they overcame the disease.

Today, most tuberculosis patients can be treated successfully with drugs that prevent the bacteria from multiplying. In the United States, the number of tuberculosis cases has been declining, but in some developing countries, where modern treatments and medications are not available, tuberculosis remains a constant threat. The World Health Organization estimated that in the 1990s, there would be approximately 90 million new cases of TB, with approximately 30 million deaths.

In Exercises 1–5, use the following information.

The number of newly reported cases of tuberculosis T (in thousands) in the United States from 1991 to 1996 can be approximated by the equation

$$T = 28.5(0.9567)^t,$$

where t represents the number of years since 1990.

1. Identify the initial amount, the decay factor, and the annual percent decrease.
2. Sketch the graph of the model.
3. In what year was the number of newly reported cases in the United States approximately 25,000?
4. Use a graphing calculator to determine the year when the number of newly reported cases will be approximately 16,000.
5. Estimate the number of newly reported cases in 2005.

NAME ______________________ DATE __________

Challenge: Skills and Applications

For use with pages 474–479

In Exercises 1–2, suppose $f(x) = ab^x$.

1. Find a and b if $f(0) = \frac{8}{3}$ and $f(3) = \frac{1}{81}$.
2. Find $f(5)$ if $f(1) = 2$ and $f(2) = \frac{4}{3}$.
3. The value of a car depreciates by a fixed percent each year.
   a. Write an equation that gives the current value V of the car in terms of the original value V_0, the fixed percent p expressed as a decimal, and time t in years.
   b. Suppose that the original value of the car was \$16,000, and after 2 years this value had depreciated to \$9000. Find the value of the car after 5 years, to the nearest dollar.
4. Carbon-14, an unstable isotope of carbon which decays exponentially to a more stable form, is used to date animal remains. After 5700 years, one-half of the original amount of carbon-14, by weight, remains.
   a. Write an exponential function with base one-half relating the amount N of carbon-14 in the animal remains after t years to the original amount N_0.
   b. Suppose an animal contained 3.2 mg of carbon-14 when it was alive. Estimate how long ago the animal died, if its remains contain 0.4 mg of carbon-14 today.
5. a. Describe the common features of all graphs of functions of the form $y = 3a^{-x}$ for all possible values of $0 < a < 1$.
   b. Repeat part (a) for functions of the form $y = -5a^{-x}$.
6. Let $f(x) = b^x$. For a fixed positive integer n, let $g(x) = [f(x)]^n$, and let $h(x) = f(f(f(\ldots(x)\ldots)$ (n compositions).
   a. Express $g(x)$ as $f(?)$. (*Hint:* Express $g(x)$ in terms of b, n and x.)
   b. Express $h(x)$ as $f(?)$.

LESSON 8.3

TEACHER'S NAME ________________ CLASS ______ ROOM ______ DATE ______

Lesson Plan

2-day lesson (See *Pacing the Chapter,* TE pages 462C–462D) **For use with pages 480–485**

GOALS

1. **Use the number *e* as the base of exponential functions.**
2. **Use the natural base *e* in real-life situations.**

State/Local Objectives ________________________________

____

✓ Check the items you wish to use for this lesson.

STARTING OPTIONS

____ Homework Check: TE page 477; Answer Transparencies
____ Warm-Up or Daily Homework Quiz: TE pages 480 and 479, CRB page 39, or Transparencies

TEACHING OPTIONS

____ Lesson Opener (Application): CRB page 40 or Transparencies
____ Examples: Day 1: 1–3, SE pages 480–481; Day 2: 4–5, SE page 482
____ Extra Examples: Day 1: TE page 481 or Transp.; Day 2: TE page 482 or Transp.
____ Closure Question: TE page 482
____ Guided Practice: SE page 483 Day 1: Exs. 1–15; Day 2: Ex. 16

APPLY/HOMEWORK

Homework Assignment

____ Basic Day 1: 18–44 even, 50–60 even, 61–66; Day 2: 45–48, 68–78 even, 81, 82, 84–94 even; Quiz 1: 1–16
____ Average Day 1: 18–60 even, 61–66, 68–74 even; Day 2: 75–78, 81–83, 84–94 even; Quiz 1: 1–16
____ Advanced Day 1: 18–60 even, 61–72; Day 2: 73–79, 81–83, 84–94 even; Quiz 1: 1–16

Reteaching the Lesson

____ Practice Masters: CRB pages 41–43 (Level A, Level B, Level C)
____ Reteaching with Practice: CRB pages 44–45 or Practice Workbook with Examples
____ Personal Student Tutor

Extending the Lesson

____ Applications (Real-Life): CRB page 47
____ Challenge: SE page 485; CRB page 48 or Internet

ASSESSMENT OPTIONS

____ Checkpoint Exercises: Day 1: TE page 481 or Transp.; Day 2: TE page 482 or Transp.
____ Daily Homework Quiz (8.3): TE page 485, CRB page 52, or Transparencies
____ Standardized Test Practice: SE page 485; TE page 485; STP Workbook; Transparencies
____ Quiz (8.1–8.3): SE page 485; CRB page 49

Notes ________________________________

____

____

LESSON

TEACHER'S NAME ______ CLASS ______ ROOM ______ DATE ______

Lesson Plan for Block Scheduling

1-day lesson (See *Pacing the Chapter,* TE pages 462C–462D) **For use with pages 480–485**

GOALS
1. Use the number *e* as the base of exponential functions.
2. Use the natural base *e* in real-life situations.

State/Local Objectives ______

CHAPTER PACING GUIDE	
Day	**Lesson**
1	8.1 (all); 8.2(all)
2	**8.3 (all)**
3	8.4 (all)
4	8.5 (all)
5	8.6 (all)
6	8.7 (all); 8.8(all)
7	Review/Assess Ch. 8

✓ Check the items you wish to use for this lesson.

STARTING OPTIONS

____ Homework Check: TE page 477; Answer Transparencies
____ Warm-Up or Daily Homework Quiz: TE pages 480 and 479, CRB page 39, or Transparencies

TEACHING OPTIONS

____ Lesson Opener (Application): CRB page 40 or Transparencies
____ Examples: 1–5: SE pages 480–482
____ Extra Examples: TE pages 481–482 or Transparencies
____ Closure Question: TE page 482
____ Guided Practice Exercises: SE page 483

APPLY/HOMEWORK

Homework Assignment

____ Block Schedule: 18–60 even, 61–66, 68–74 even, 75–78, 81–83, 84–94 even; Quiz 1: 1–16

Reteaching the Lesson

____ Practice Masters: CRB pages 41–43 (Level A, Level B, Level C)
____ Reteaching with Practice: CRB pages 44–45 or Practice Workbook with Examples
____ Personal Student Tutor

Extending the Lesson

____ Applications (Real Life): CRB page 47
____ Challenge: SE page 485; CRB page 48 or Internet

ASSESSMENT OPTIONS

____ Checkpoint Exercises: TE pages 481–482 or Transparencies
____ Daily Homework Quiz (8.3): TE page 485, CRB page 52, or Transparencies
____ Standardized Test Practice: SE page 485; TE page 485; STP Workbook; Transparencies
____ Quiz (8.1–8.3): SE page 485; CRB page 49

Notes ______

NAME ______________________ DATE ________

Available as a transparency

Warm-Up Exercises

For use before Lesson 8.3, pages 480–485

Simplify. Round to the nearest hundredth.

1. $\left(1 + \frac{1}{2}\right)^2$

2. $\left(1 + \frac{1}{3}\right)^3$

3. $\left(1 + \frac{1}{4}\right)^4$

4. $\left(1 + \frac{1}{5}\right)^5$

State the domain and range of the function.

5. $y = 3\left(\frac{1}{2}\right)^x$

6. $y = -3(2)^x$

Daily Homework Quiz

For use after Lesson 8.2, pages 473–479

Tell whether the function represents *exponential growth* or *exponential decay*.

1. $f(x) = 0.2 \cdot 1.5^x$

2. $f(x) = 10 \cdot 3^{-x}$

3. Graph $y = 3 \cdot 0.5^x$.

4. Graph $y = -2\left(\frac{1}{4}\right)^{x+2} + 2$.

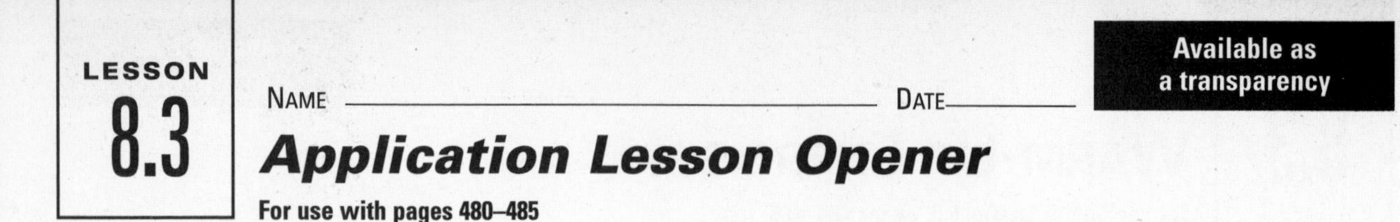

Application Lesson Opener

For use with pages 480–485

Suppose you live in a country where the rate of inflation is so great that savings accounts are offered at an interest rate of 100% per year, compounded n times per year, where n is allowed to vary.

Suppose you invest $1,000.00 in a savings account.

1. Use the compound interest formula, $A = P\left(1 + \frac{r}{n}\right)^{nt}$, to find the amount of money in the account after 1 year. Your answer should be an equation showing A as a function of n.

2. Complete the table. (Use a calculator.)

Compounding	*n*	*Amount after 1 year (dollars)*
Annually	1	
Quarterly	4	
Monthly	12	
Daily	365	
Hourly	8760	
Every minute	525,600	

3. As n increases, the situation approaches what banks call *continuous compounding*. Try several larger values of n to guess how much money would be in the account after 1 year under continuous compounding.

Lesson 8.3

NAME ______________________ DATE __________

Practice A

For use with pages 480–485

Use a calculator to evaluate the expression. Round the result to three decimal places.

1. e^4
2. e^{-1}
3. e^7
4. e^0
5. e^{-2}
6. $e^{2/3}$
7. $e^{-1/2}$
8. $e^{2.3}$

Tell whether the function is an example of *exponential growth* or *exponential decay*.

9. $f(x) = e^x$
10. $f(x) = e^{-x}$
11. $f(x) = 2e^x$
12. $f(x) = \frac{1}{2}e^x$
13. $f(x) = e^{-2x}$
14. $f(x) = e^{-1/3x}$

Simplify the expression.

15. $e^3 \cdot e^5$
16. $e^{-2} \cdot e^8$
17. $(e^2)^5$
18. $\frac{e^8}{e^5}$
19. $\frac{e^{-3}}{e^2}$
20. $(2e^5)^3$

Match the function with its graph.

21. $f(x) = 2e^x + 1$
22. $f(x) = 2e^{x+1}$
23. $f(x) = e^{2x}$

A.

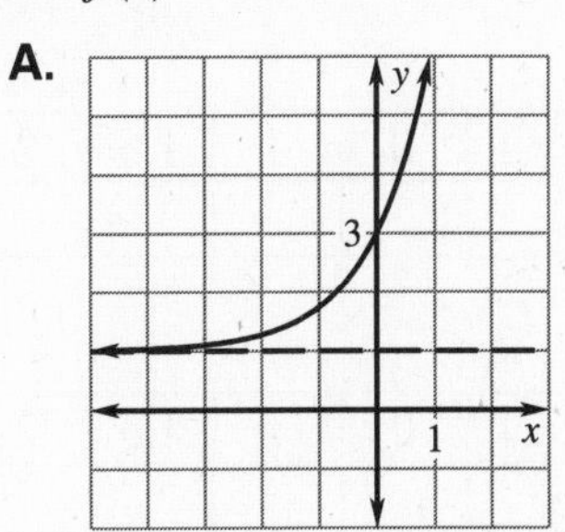

B.

C.

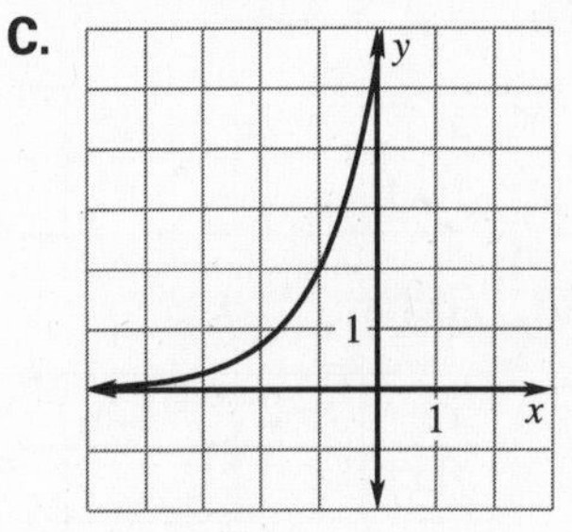

24. ***Continuous Compounding*** You deposit \$725 in an account that pays 4.5% annual interest compounded continuously. What is the balance after 3 years?

25. ***Population*** The population P of a city can be modeled by $P = 250{,}000e^{0.01t}$ where t is the number of years since 1990. What was the population in 1999?

LESSON 8.3

NAME ______________________ DATE __________

Practice B

For use with pages 480–485

Use a calculator to evaluate the expression. Round the result to three decimal places.

1. e^5
2. $e^{-1/3}$
3. $e^{-1.4}$
4. $e^{\sqrt{2}}$

Tell whether the function is an example of *exponential growth* or *exponential decay*.

5. $f(x) = 2e^{3x}$
6. $f(x) = e^{-3x}$
7. $f(x) = 2e^{-3x}$
8. $f(x) = \frac{1}{5}e^{5x}$
9. $f(x) = \frac{1}{2}e^{-x}$
10. $f(x) = 4e^{5x}$

Simplify the expression.

11. $(e^4)^{-2}$
12. $\dfrac{3e^5}{e}$
13. $\left(\dfrac{e}{2}\right)^{-1}$
14. $(4e^3)^2$
15. $-3e \cdot 4e^2$
16. $2e^x \cdot e^{x+3}$
17. $\sqrt{64e^{4x}}$
18. $e^{2x} \cdot e^{1-2x}$
19. $\dfrac{e}{e^{x+1}}$

Complete the table of values. Round to two decimal places.

20. $f(x) = 2e^x$

x	-2	-1.5	-1	0	1	1.5	2
$f(x)$							

21. $f(x) = 2e^{-x}$

x	-2	-1.5	-1	0	1	1.5	2
$f(x)$							

22. $f(x) = e^{2x} + 3$

x	-2	-1.5	-1	0	1	1.5	2
$f(x)$							

23. $f(x) = e^{-3x} - 2$

x	-2	-1.5	-1	0	1	1.5	2
$f(x)$							

Graph the function and identify the horizontal asymptote.

24. $f(x) = 2e^x$
25. $f(x) = 2e^{-x}$
26. $f(x) = e^x + 2$
27. $f(x) = e^{-3x} + 1$
28. $f(x) = \frac{1}{2}e^{2x} - 1$
29. $f(x) = e^{-2.5x} - 3$

Interest **In Exercises 30–32, use the following information.**

You deposit \$1200 in an account that pays 5% annual interest. After 10 years, you withdraw the money.

30. Find the balance in the account if the interest was compounded quarterly.
31. Find the balance in the account if the interest was compounded continuously.
32. Which type of compounding yielded the greatest balance?

Lesson 8.3

NAME ______________________ DATE __________

Practice C

For use with pages 480–485

Use a calculator to evaluate the expression. Round the result to three decimal places.

1. $e^{\sqrt{3}}$
2. $e^{-2.6}$
3. $e^{-\frac{1}{\sqrt{2}}}$
4. e^e

Simplify the expression.

5. $e^2(2e^4)^3$
6. $\left(\frac{1}{3}e^{-2}\right)^{-4}$
7. $\left(\frac{e^2}{2}\right)^{-3}$
8. $(4e^{0.5x})^{-6}$
9. $\left(\frac{e^{3x}}{2e}\right)^2$
10. $\sqrt[3]{8e^{12x}}$

Identify the horizontal asymptote of the function.

11. $f(x) = 3e^{2x} - 1$
12. $f(x) = \frac{1}{2}e^{3x+1} + 4$
13. $f(x) = 245e^{-0.023x}$

Graph the function. State the domain and range.

14. $f(x) = 2e^{3x} + 1$
15. $f(x) = \frac{1}{4}e^{-x} - 2$
16. $f(x) = 2e^{x-4} + 1$
17. $f(x) = \frac{1}{2}e^{2x+1} - 5$
18. $f(x) = \frac{2}{3}e^{3-x} + 1$
19. $f(x) = \frac{5}{4}e^{2(x-1)} - 3$

Carbon Dating **In Exercises 20–22, use the following information.**

Carbon dating is a process to estimate the age of organic material. In carbon dating the formula used is

$$R = \frac{1}{10^{12}}e^{-t/8233}$$

where R is the ratio of Carbon 14 to Carbon 12 and t is time in years.

20. Is the model an example of exponential growth or exponential decay?
21. Graph the function.
22. Use the graph to estimate the age of a fossil whose Carbon 14 to Carbon 12 ratio is 3×10^{-13}.

Learning Curve **In Exercises 23–26, use the following information.**

The management at a factory has determined that a worker can produce a maximum of 30 units per day. The model $y = 30 - 30e^{-0.07t}$ indicates the number of units y that a new employee can produce per day after t days on the job.

23. Is the model an example of exponential growth or exponential decay?
24. Graph the function.
25. How many units can be produced per day by an employee who has been on the job 8 days?
26. Use the graph to estimate how many days of employment are required for a worker to produce 25 units per day.

LESSON 8.3

NAME ______________________ DATE ________

Reteaching with Practice

For use with pages 480–485

GOAL **Use the number *e* as the base of exponential functions**

VOCABULARY

The **natural base *e*** is irrational. It is defined as follows:

As n approaches $+\infty$, $\left(1 + \frac{1}{n}\right)^n$ approaches $e \approx 2.718281828459$.

EXAMPLE 1 *Simplifying Natural Base Expressions*

Simplify the expression.

a. $2e \cdot e^{-4}$ **b.** $\frac{6e^{5x}}{2e^{3x}}$ **c.** $(-5e^2)^3$

SOLUTION

a. $2e \cdot e^{-4} = 2e^{1+(-4)}$
$= 2e^{-3}$
$= \frac{2}{e^3}$

b. $\frac{6e^{5x}}{2e^{3x}} = 3e^{5x-3x}$
$= 3e^{2x}$

c. $(-5e^2)^3 = (-5)^3e^{(2)(3)}$
$= -125e^6$

Exercises for Example 1

Simplify the expression.

1. $e^{-2} \cdot e^6$ **2.** $5e^3 \cdot 4e^2$ **3.** $e^{2x} \cdot e^{4x}$

4. $(2e^3)^3$ **5.** $\frac{e^5}{e^2}$ **6.** $\frac{10e^2}{2e^4}$

Lesson 8.3

EXAMPLE 2 *Evaluating Natural Base Expressions*

Use a calculator to evaluate the expression (a) $e^{2/3}$ and (b) e^{-2}.

SOLUTION

	Expression	*Keystrokes*	*Display*
a.	$e^{2/3}$	2nd [e^x] 2 ÷ 3) ENTER	1.947734041
b.	e^{-2}	2nd [e^x] (-) 2) ENTER	0.1353352832

Exercises for Example 2

Use a calculator to evaluate the expression. Round the result to three decimal places.

7. e^4 **8.** $e^{1/3}$ **9.** $e^{1.2}$ **10.** $2e^{-1/5}$

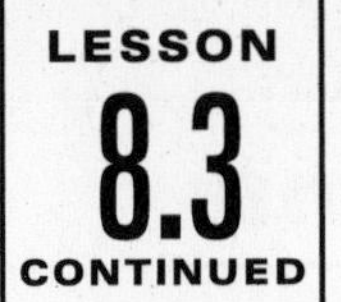

NAME ______________________ DATE __________

Reteaching with Practice

For use with pages 480–485

EXAMPLE 3 Graphing Natural Base Functions

Graph the function. State the domain and range.

a. $y = 3e^{-2x}$

b. $y = \frac{1}{2}e^{x} - 5$

SOLUTION

a. Because $a = 3$ is positive and $r = -2$ is negative, the function is an exponential decay function. Plot points $(0, 3)$ and $(1, 0.41)$ and draw the curve.

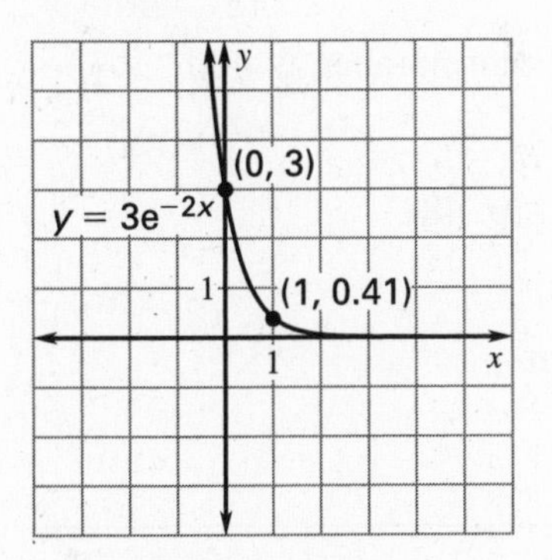

The domain is all real numbers, and the range is $y > 0$.

b. Because $a = \frac{1}{2}$ is positive and $r = 1$ is positive, the function is an exponential growth function. Translate the graph of $y = \frac{1}{2}e^{x}$ down 5 units.

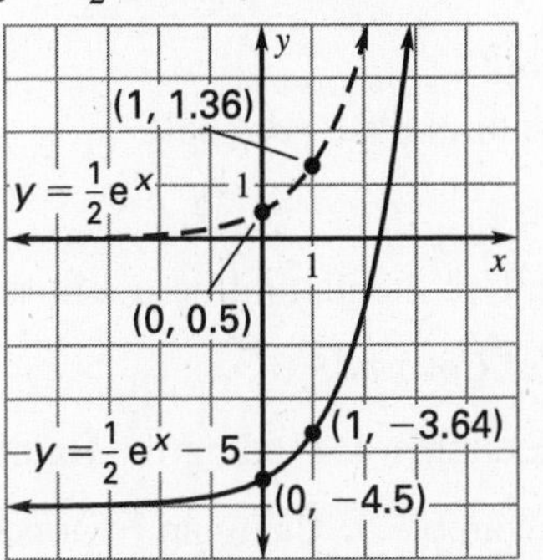

The domain is all real numbers, and the range is $y > -5$.

Exercises for Example 3

Graph the function. State the domain and range.

11. $y = 2e^{-x}$

12. $y = e^{x-3}$

13. $y = 4e^{x} - 3$

14. $y = e^{-2x} + 1$

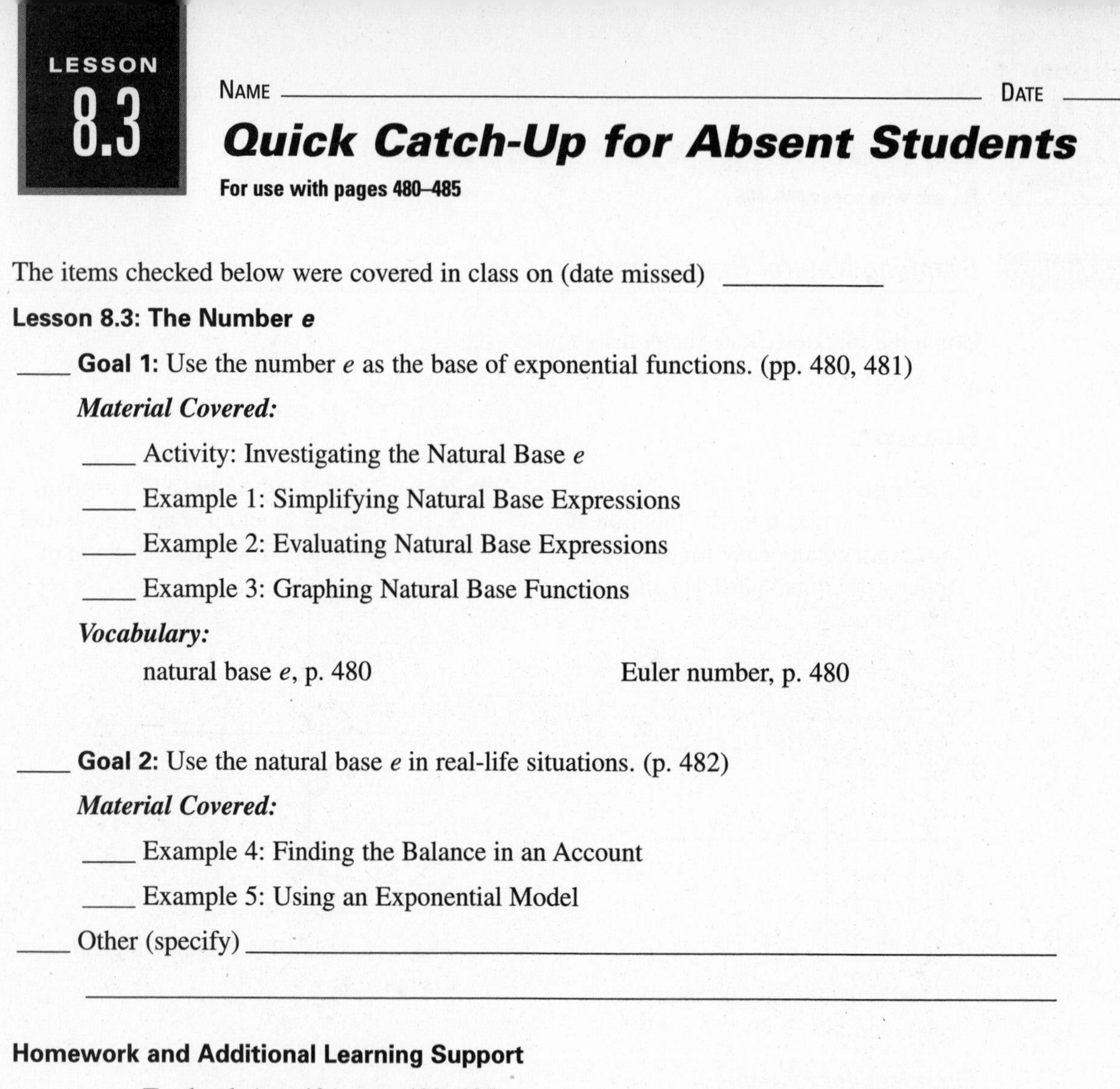

LESSON 8.3

NAME ____ DATE __________

Quick Catch-Up for Absent Students

For use with pages 480–485

The items checked below were covered in class on (date missed) ____________

Lesson 8.3: The Number *e*

____ **Goal 1:** Use the number *e* as the base of exponential functions. (pp. 480, 481)

Material Covered:

____ Activity: Investigating the Natural Base *e*

____ Example 1: Simplifying Natural Base Expressions

____ Example 2: Evaluating Natural Base Expressions

____ Example 3: Graphing Natural Base Functions

Vocabulary:

natural base *e*, p. 480 Euler number, p. 480

____ **Goal 2:** Use the natural base *e* in real-life situations. (p. 482)

Material Covered:

____ Example 4: Finding the Balance in an Account

____ Example 5: Using an Exponential Model

____ Other (specify) ____

____

Homework and Additional Learning Support

____ Textbook (specify) pp. 483–485

____

____ *Reteaching with Practice* worksheet (specify exercises) ____________________

____ *Personal Student Tutor* for Lesson 8.3

Lesson 8.3

NAME ______________________________ DATE ____________

Real–Life Application: When Will I Ever Use This?

For use with pages 480–485

Teachers

With the large number of people employed in education, teachers represent a significant percent of today's work force. The teaching profession itself has developed greatly since the early 1800s when the first teacher-training schools were started in Europe. Before this time, teachers received little or even no training. Today, teacher-training programs are offered at many colleges and universities, and in general, consist of three areas of study:

1. liberal arts courses,
2. advanced courses in a particular area of interest, such as mathematics, and
3. professional education courses.

Professional education courses can include courses such as teaching methods, child development, and actual classroom teaching.

In Exercises 1–5, use the following information.

There are more than 48 million teachers worldwide, with approximately four million alone in the United States. The number of public elementary and secondary school teachers T (in thousands) in the United States from 1985 to 1997 can be modeled by the equation

$$T = 2094 + 16.2x + 1.01x^2 - 3806.892e^{-x},$$

where x represents the number of years since 1980.

1. Use a graphing calculator to graph the model. Has the number of public elementary and secondary school teachers been increasing or decreasing?
2. Use the *Trace* feature of your graphing calculator to estimate when the number of public elementary and secondary teachers was about 2.25 million.
3. Use the *Trace* feature of your graphing calculator to estimate when the number of public elementary and secondary teachers was about 2.4 million.
4. Do you think this model could be used to estimate the number of public elementary and secondary school teachers in the future? Explain.
5. According to the model, when will the number of public elementary and secondary school teachers be about 3 million?

NAME _______________ DATE _______

Challenge: Skills and Applications

For use with pages 480–485

1. **a.** Let $r =$ the *annual* interest rate paid on a savings account. Suppose interest is compounded k times per year on an initial amount P in the account. Write a formula that gives the amount A in the account after t years. (Note that this means interest is compounded a total of kt times during t years.)

   b. Use the substitution $k = nr$ to eliminate k in the formula you wrote in part (a). Rewrite the formula using the variable n, but without k.

   c. What happens to the variable n that you introduced in part (b), as $k \to \infty$? Explain how the formula you wrote in part (b) reduces to Pe^{rt}, as $k \to \infty$.

In Exercises 2 and 3, use the following information: The function e^x can be approximated by the following *power series:*

$$e^x \approx 1 + x + \frac{x^2}{2!} + \frac{x^3}{3!} + \frac{x^4}{4!} + \ldots,$$

where $n!$ means $1 \cdot 2 \cdot 3 \cdot \cdot \cdot n$.

2. Use the first 3 terms of the power series for e^a and e^b to show that the first 3 terms of e^{a+b} are formed from the terms of degree ≤ 2 of the product of these two series.

3. Find the power series expression of e^{-x}. Using terms of degree ≤ 3, show that the product of the series for e^x and e^{-x} is 1.

4. In this exercise, you will prove that e is irrational, using the power series above with $x = 1$. Suppose that e is rational; that is, suppose

$$e = \frac{p}{q},$$

   with p, q positive integers.

   a. Explain why

$$(q!)e = (q!)\frac{p}{q}$$

   is an integer, as is $q!$ times the sum of the first $q + 1$ terms of the above expansion for e^x, with $x = 1$.

   b. Suppose the second number mentioned in part (a) is subtracted from the first. Explain why this difference d must be an integer.

   c. Simplify the difference d mentioned in part (b), which is given by

$$d = (q!)\left(\frac{1}{(q+1)!} + \frac{1}{(q+2)!} + \frac{1}{(q+3)!} + \ldots\right).$$

   d. Explain why $d < \frac{1}{2} + \frac{1}{4} + \frac{1}{8} + \ldots \leq 1$. How does this prove the assertion that e is irrational? Explain why $d < \frac{1}{2} + \frac{1}{4} + \frac{1}{8} + \cdot \cdot \cdot$. In Chapter 11 you will learn that this sum is equal to 1. Explain how this proves the assertion that e is irrational.

LESSON 8.3

NAME ______________________ DATE __________

Quiz 1

For use after Lessons 8.1–8.3

Graph the function. State the domain and range. *(Lessons 8.1 and 8.2)*

1. $y = 3^x + 2$

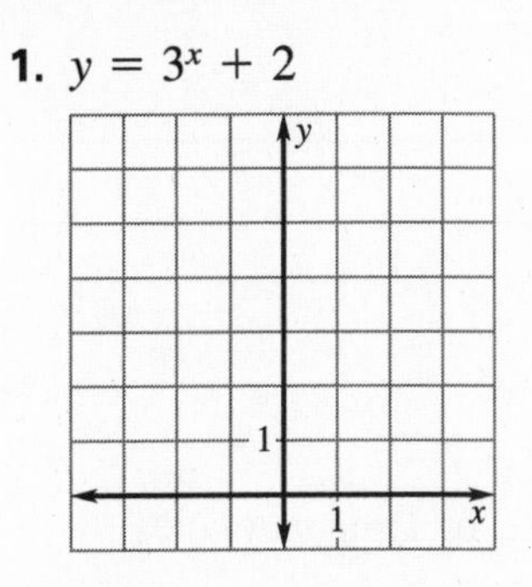

2. $y = 3\left(\frac{1}{3}\right)^x + 2$

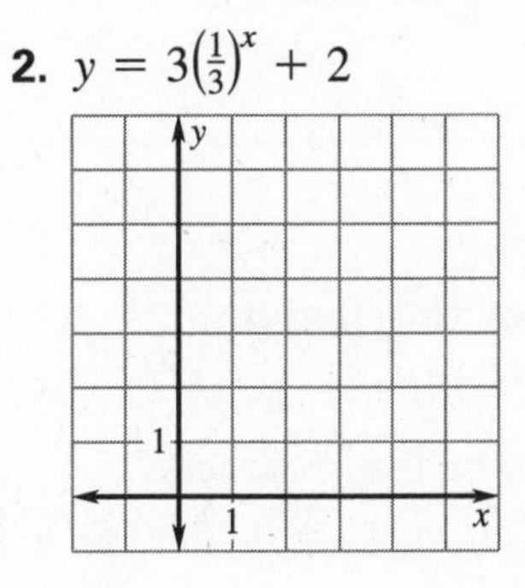

Simplify the expression. *(Lesson 8.3)*

3. $4e^{-3} \cdot e^5$

4. $(-2e^{2x})^2$

5. $(4e^{-2})^{-3x}$

6. $\dfrac{5e^x}{6e}$

7. $\dfrac{12e^x}{e^{4x}}$

8. $\sqrt[3]{27e^{6x}}$

9. Graph the function $f(x) = -5e^{2x}$. *(Lesson 8.3)*

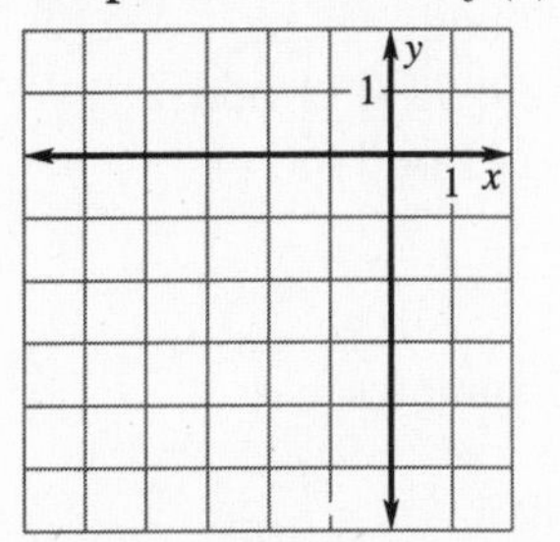

10. ***Account Balance*** You deposit $5000 in an account that pays 7% annual interest compounded continuously. What is the balance after 2 years? *(Lesson 8.3)*

Answers

1. Use grid at left.
2. Use grid at left.
3. ______
4. ______
5. ______
6. ______
7. ______
8. ______
9. Use grid at left.
10. ______

Lesson 8.3

TEACHER'S NAME ______________________ CLASS ______ ROOM ______ DATE ______

Lesson Plan

2-day lesson (See *Pacing the Chapter,* TE pages 462C–462D) For use with pages 486–492

GOALS
1. **Evaluate logarithmic functions.**
2. **Graph logarithmic functions.**

State/Local Objectives ____

____

✓ Check the items you wish to use for this lesson.

STARTING OPTIONS
____ Homework Check: TE page 483; Answer Transparencies
____ Warm-Up or Daily Homework Quiz: TE pages 486 and 485, CRB page 52, or Transparencies

TEACHING OPTIONS
____ Motivating the Lesson: TE page 487
____ Lesson Opener (Activity): CRB page 53 or Transparencies
____ Graphing Calculator Activity with Keystrokes: CRB pages 54–55
____ Examples: Day 1: 1–4, SE pages 486–487; Day 2: 5–8, SE pages 488–489
____ Extra Examples: Day 1: TE page 487 or Transp.; Day 2: TE pages 488–489 or Transp.; Internet
____ Closure Question: TE page 489
____ Guided Practice: SE page 490 Day 1: Exs. 1–12; Day 2: Exs. 13–15

APPLY/HOMEWORK
Homework Assignment
____ Basic Day 1: 16–46 even, 77, 78; Day 2: 48–76 even, 80, 82–87, 93–107 odd
____ Average Day 1: 16–34 even, 36–47, 77–79; Day 2: 48–76 even, 80, 82–89, 93–111 odd
____ Advanced Day 1: 16–34 even, 36–47, 77–79; Day 2: 48–74, 80–92, 93–111 odd

Reteaching the Lesson
____ Practice Masters: CRB pages 56–58 (Level A, Level B, Level C)
____ Reteaching with Practice: CRB pages 59–60 or Practice Workbook with Examples
____ Personal Student Tutor

Extending the Lesson
____ Cooperative Learning Activity: CRB page 62
____ Applications (Interdisciplinary): CRB page 63
____ Challenge: SE page 492; CRB page 64 or Internet

ASSESSMENT OPTIONS
____ Checkpoint Exercises: Day 1: TE page 487 or Transp.; Day 2: TE pages 488–489 or Transp.
____ Daily Homework Quiz (8.4): TE page 492, CRB page 67, or Transparencies
____ Standardized Test Practice: SE page 492; TE page 492; STP Workbook; Transparencies

Notes ____

____

____

LESSON
8.4

TEACHER'S NAME ______________________ CLASS ______ ROOM ______ DATE ______

Lesson Plan for Block Scheduling

1-day lesson (See *Pacing the Chapter,* TE pages 462C–462D) **For use with pages 486–492**

1. Evaluate logarithmic functions.
2. Graph logarithmic functions.

State/Local Objectives ______________________

____

____

CHAPTER PACING GUIDE	
Day	**Lesson**
1	8.1 (all); 8.2(all)
2	8.3 (all)
3	**8.4 (all)**
4	8.5 (all)
5	8.6 (all)
6	8.7 (all); 8.8(all)
7	Review/Assess Ch. 8

✓ Check the items you wish to use for this lesson.

STARTING OPTIONS

____ Homework Check: TE page 483; Answer Transparencies
____ Warm-Up or Daily Homework Quiz: TE pages 486 and 485, CRB page 52, or Transparencies

TEACHING OPTIONS

____ Motivating the Lesson: TE page 487
____ Lesson Opener (Activity): CRB page 53 or Transparencies
____ Graphing Calculator Activity with Keystrokes: CRB pages 54–55
____ Examples: 1–8: SE pages 486–489
____ Extra Examples: TE pages 487–489 or Transparencies; Internet
____ Closure Question: TE page 489
____ Guided Practice Exercises: SE page 490

APPLY/HOMEWORK

Homework Assignment

____ Block Schedule: 16–34 even, 36–47, 48–76 even, 77–80, 82–89, 93–111 odd

Reteaching the Lesson

____ Practice Masters: CRB pages 56–58 (Level A, Level B, Level C)
____ Reteaching with Practice: CRB pages 59–60 or Practice Workbook with Examples
____ Personal Student Tutor

Extending the Lesson

____ Cooperative Learning Activity: CRB page 62
____ Applications (Interdisciplinary): CRB page 63
____ Challenge: SE page 492; CRB page 64 or Internet

ASSESSMENT OPTIONS

____ Checkpoint Exercises: TE pages 487–489 or Transparencies
____ Daily Homework Quiz (8.4): TE page 492, CRB page 67, or Transparencies
____ Standardized Test Practice: SE page 492; TE page 492; STP Workbook; Transparencies

Notes ______________________

____

____

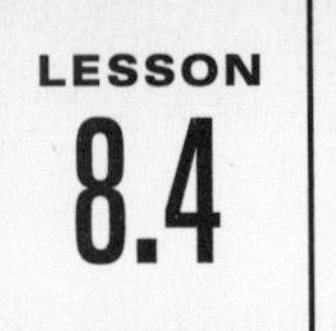

LESSON 8.4

NAME ______________________ DATE __________

Available as a transparency

WARM-UP EXERCISES

For use before Lesson 8.4, pages 486–492

Find the value of *x*.

1. $3^x = 9$

2. $x^3 = -8$

3. $10^0 = x$

4. $10^x = 0.001$

5. $\left(\frac{3}{2}\right)^{-1} = x$

DAILY HOMEWORK QUIZ

For use after Lesson 8.3, pages 480–485

Simplify the expression.

1. $2e^{2x} \cdot e^x$

2. $\dfrac{6e^{-x}}{12e^{3x}}$

Tell whether the function represents *exponential growth* or *exponential decay*.

3. $f(x) = 0.1 \cdot e^{0.3x}$

4. $f(x) = 5 \cdot e^{-4x}$

5. Graph $y = 0.2e^{2x+1} - 2$. State the domain and range.

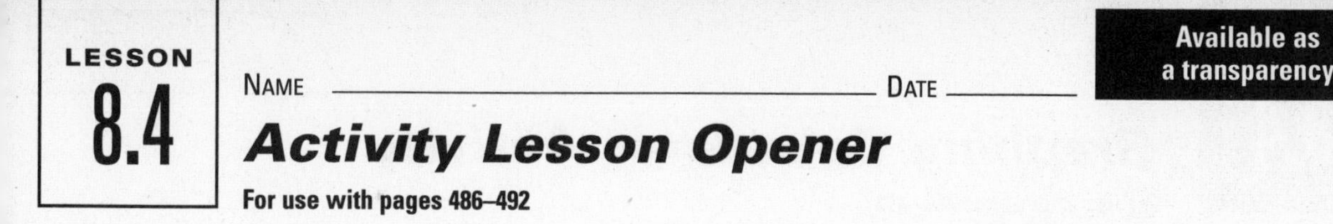

Activity Lesson Opener

For use with pages 486–492

SET UP: Work in groups of 2 or 3.

YOU WILL NEED: • index cards • pencil and paper

If b and y are positive numbers $(b \neq 1)$, the *logarithm* of *y with base b* (or, "log base b of y") is denoted $\log_b y$ and is defined as follows:

$$\log_b y = x \text{ if and only if } b^x = y$$

Thus, to find $\log_3 81$, determine how many 3's you would need to multiply to get 81: $\log_3 81 = 4$ because $3^4 = 81$.

Follow the steps below to play a game involving logarithms:

- Write the following expressions on index cards (one per card) to create a set of 30 Logarithm Cards.

$\log_2 1$	$\log_2 2$	$\log_2 4$	$\log_2 8$	$\log_2 16$	$\log_2 32$
$\log_3 1$	$\log_3 3$	$\log_3 9$	$\log_3 27$	$\log_3 81$	$\log_3 243$
$\log_4 1$	$\log_4 4$	$\log_4 16$	$\log_4 64$	$\log_4 256$	$\log_4 1024$
$\log_5 1$	$\log_5 5$	$\log_5 25$	$\log_6 6$	$\log_6 36$	$\log_6 216$
$\log_7 7$	$\log_7 49$	$\log_{10} 1$	$\log_{10} 10$	$\log_{10} 100$	$\log_{10} 1000$

- Take turns drawing cards. Work with the students in your group to determine the value of the expression on your card. The value of the expression is the number of points you receive for the turn.

- The winner is the first player to obtain a total score of 25 points or more.

Note: As an alternative to using index cards, you can use scissors to cut the logarithm expressions out from a copy of this page. The small slips of paper can be placed in a hat and drawn at random.

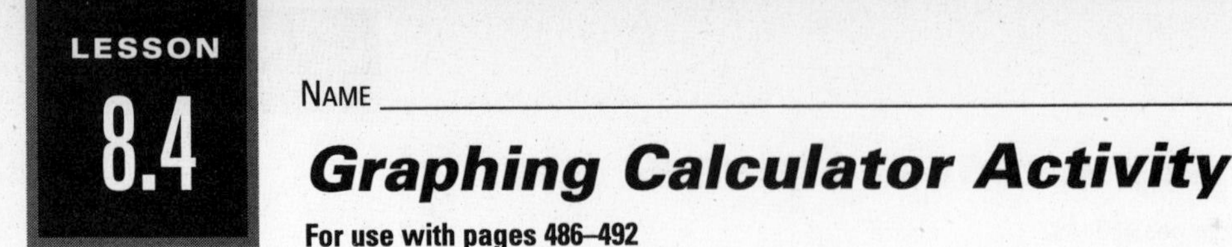

LESSON 8.4

NAME ______________________ DATE ________

Graphing Calculator Activity

For use with pages 486–492

GOAL **To discover the inverse of a logarithmic function**

Let b and y be positive numbers, $b \neq 1$. The logarithm of y with base b is denoted by $\log_b y$ and is defined as follows:

$$\log_b y = x \text{ if and only if } b^x = y.$$

Activity

1. Graph $y = \log_{10} x$.
2. Graph the inverse of $y = \log_{10} x$.
3. Graph $y = \log_{10} x$ and $y = 10^x$.
4. Trace the two functions to verify that they are inverses.

Exercises

1. The number e is approximately equal to 2.718. It is seen often in scientific equations. The logarithm with base e can be denoted by $\log_e$, but it is more often denoted by ln. Use your graphing calculator to graph $y = \ln x$ and $y = e^x$.
2. Use the trace functions to verify that $y = \ln x$ and $y = e^x$ are inverses.
3. Use a graphing calculator to graph $y = 5^x$. As you did in Step 2, use the inverse function on your calculator to graph the inverse of $y = 5^x$. This will be $y = \log_5 x$.
4. Most calculators only graph logs with a base of 10 or e. However, $y = \log_5 x$ can be written as $y = (\log_{10} x)/(\log_{10} 5)$ or as $y = (\ln x)/(\ln 5)$. Use this idea to graph $y = \log_5 x$. Does it have the same graph as the inverse of $y = 5^x$?
5. Use a graphing calculator to graph $(0.5)^x$. Use the inverse function on your calculator to graph the inverse of $y = (0.5)^x$. This will be $y = \log_{0.5} x$.
6. Use the idea in Exercise 4 to graph $y = \log_{0.5} x$ directly.
7. Use a graphing calculator to graph $y = (8.5)^x$ and $y = \log_{8.5} x$.
8. Use a graphing calculator to graph $y = \log_{10}(x + 2)$ and find the inverse.
9. Use a graphing calculator to graph $y = (\log_{10} x) + 5$ and find the inverse.

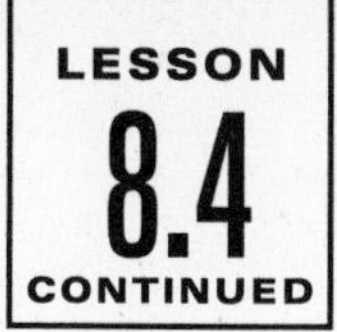

LESSON 8.4 CONTINUED

NAME ______________________ DATE ____________

Graphing Calculator Activity

For use with pages 486–492

TI-82

Step 1

Y= **LOG** **X,T,θ** **WINDOW** **ENTER** **(-)** 3

ENTER 3 **ENTER** 1 **ENTER** **(-)** 3 **ENTER**

3 **ENTER** 1 **ENTER** **GRAPH**

Step 2

2nd [DRAW] 8

2nd [Y-VARS] 1 1 **ENTER**

Step 3

Y= **▼** 10 **^** **X,T,θ** **ENTER** **GRAPH**

Step 4

Press **TRACE** and then use the cursor keys to see that when (a, b) is a solution of $y = \log_{10} x$ that (b, a) is a solution of the inverse. For example, $(1, 0)$ is a solution of $y = \log_{10} x$.

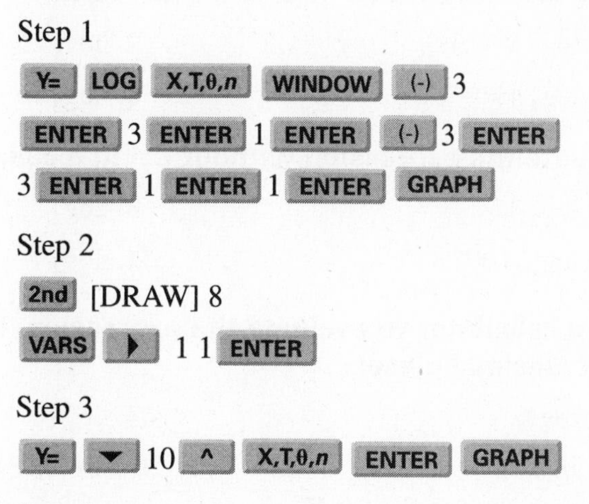

TI-83

Step 1

Y= **LOG** **X,T,θ,n** **WINDOW** **(-)** 3

ENTER 3 **ENTER** 1 **ENTER** **(-)** 3 **ENTER**

3 **ENTER** 1 **ENTER** 1 **ENTER** **GRAPH**

Step 2

2nd [DRAW] 8

VARS **▶** 1 1 **ENTER**

Step 3

Y= **▼** 10 **^** **X,T,θ,n** **ENTER** **GRAPH**

Step 4

Press **TRACE** and then use the cursor keys to see that when (a, b) is a solution of $y = \log_{10} x$ that (b, a) is a solution of the inverse. For example, $(1, 0)$ is a solution of $y = \log_{10} x$.

SHARP EL-9600c

Step 1

Y= **log** **X/θ/T/n** **WINDOW** **(-)** 3

ENTER 3 **ENTER** 1 **ENTER** **(-)** 3 **ENTER**

3 **ENTER** 1 **ENTER** **GRAPH**

Step 2

2ndF [DRAW] [A] 8

VARS [A] **ENTER** [A] 1 **ENTER**

Step 3

Y= **▼** 10 **a^b** **X/θ/T/n** **ENTER** **GRAPH**

Step 4

Press **TRACE** and then use the cursor keys to see that when (a, b) is a solution of $y = \log_{10} x$ that (b, a) is a solution of the inverse. For example, $(1, 0)$ is a solution of $y = \log_{10} x$.

CASIO CFX-9850GA PLUS

From the main menu, choose GRAPH.

Step 1

log **X,θ,T** **EXE** **SHIFT** **F3** **(-)** 3

EXE 3 **EXE** 1 **EXE** **(-)** 3 **EXE** 3 **EXE** 1

EXE **EXIT** **F6**

Step 2

SHIFT **F4** **F4**

Step 3

SHIFT [QUIT] **▼** 10 **^** **X,θ,T** **EXE**

GRAPH

Step 4

F1

Use the cursor keys to see that when (a, b) is a solution of $y = \log_{10} x$ that (b, a) is a solution of the inverse. For example, $(1, 0)$ is a solution of $y = \log_{10} x$.

NAME ______________________ DATE ________

Practice A

For use with pages 486–492

Rewrite the equation in exponential form.

1. $\log_2 8 = 3$
2. $\log_5 25 = 2$
3. $\log_3 27 = 3$
4. $\log_7 49 = 2$
5. $\log_2 16 = 4$
6. $\log_6 6 = 1$

Evaluate the expression without using a calculator.

7. $\log_2 4$
8. $\log_2 32$
9. $\log_8 64$
10. $\log_{10} 100$
11. $\log_7 1$
12. $\log_8 8$

Use a calculator to evaluate the expression. Round the result to three decimal places.

13. $\log 6$
14. $\log (0.4)$
15. $\log 3.72$
16. $\ln 8$
17. $\ln (0.23)$
18. $\ln (6.12)$

Simplify the expression.

19. $7^{\log_7 x}$
20. $27^{\log_{27} x}$
21. $13^{\log_{13} x}$
22. $\log_3(3^x)$
23. $\log_{15}(15^x)$
24. $\log_{221}(221^x)$

Match the function with its graph.

25. $f(x) = \log_3 x$
26. $f(x) = \log_5 x$
27. $f(x) = \log_{1/2} x$

A.

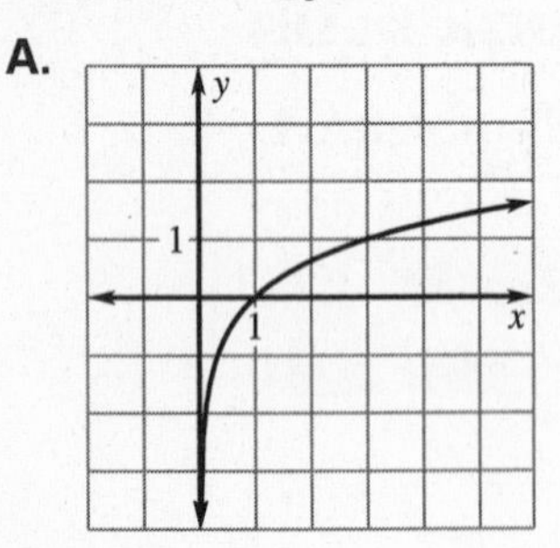

B.

C.

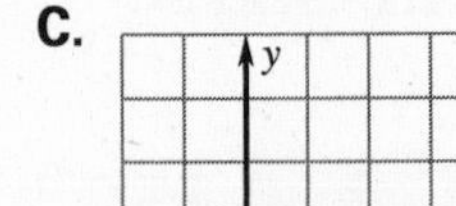

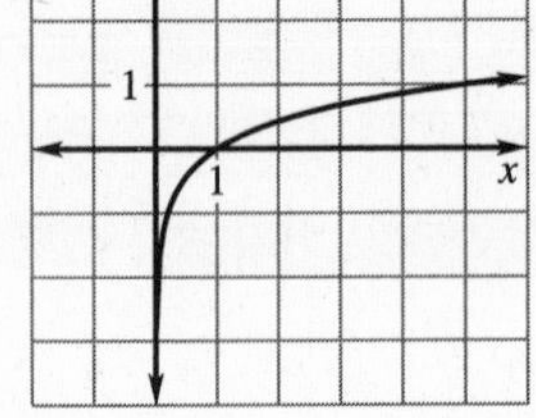

Match the function with the graph of its inverse.

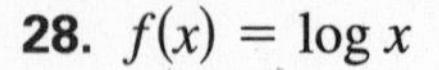

28. $f(x) = \log x$

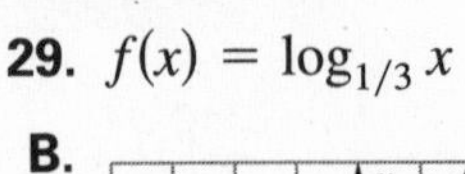

29. $f(x) = \log_{1/3} x$

30. $f(x) = \ln x$

A.

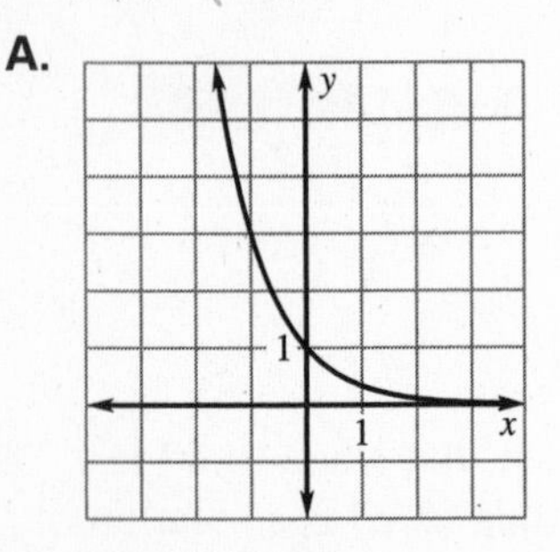

B.

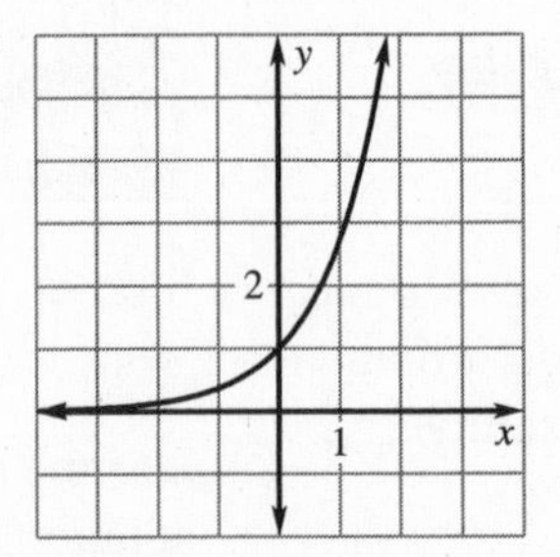

C.

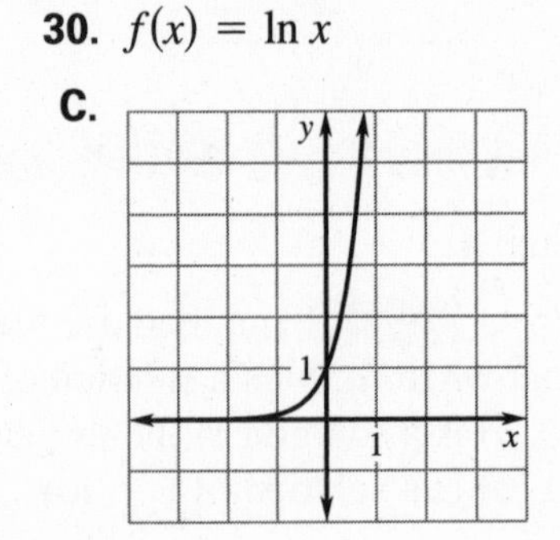

31. *Sound* The level of sound V in decibels with an intensity I can be modeled by

$$V = 10 \log\left(\frac{I}{10^{-16}}\right),$$

where I is intensity in watts per centimeter. Loud music can have an intensity of 10^{-5} watts per centimeter. Find the level of sound of loud music.

LESSON 8.4

NAME ______________________ DATE ______

Practice B

For use with pages 486–492

Rewrite the equation in exponential form.

1. $\log_4 16 = 2$
2. $\log_3 81 = 4$
3. $\log_2 1 = 0$
4. $\log_9 3 = \frac{1}{2}$
5. $\log_5 \frac{1}{5} = -1$
6. $\log_2 \frac{1}{8} = -3$

Use a calculator to evaluate the expression. Round the result to three decimal places.

7. $\ln \sqrt{3}$
8. $\log 11.5$
9. $\ln \left(\frac{2}{3}\right)$

Evaluate the logarithm without using a calculator.

10. $\log_3 27$
11. $\log_4 1$
12. $\log_2 \frac{1}{2}$
13. $\log_8 2$
14. $\log_5 5^{2/3}$
15. $\log_6 (-1)$

Find the inverse of the function.

16. $f(x) = \log_3 x$
17. $f(x) = \ln x$
18. $f(x) = \log_{1/3} x$
19. $f(x) = \log 2x$
20. $f(x) = \log_2 (x - 1)$
21. $f(x) = \log_4 16x$

Graph the function.

22. $f(x) = \log_6 x$
23. $f(x) = 1 + \log_6 x$
24. $f(x) = \log_6 (x + 1)$
25. $f(x) = -\log_6 x$
26. $f(x) = \log_6 (2x)$
27. $f(x) = -1 + \log_6 x$

28. ***Galloping Speed*** Four-legged animals run with two different types of motion: trotting and galloping. An animal that is trotting has at least one foot on the ground at all times. An animal that is galloping has all four feet off the ground at times. The number S of strides per minute at which an animal breaks from a trot to a gallop is related to the animal's weight w (in pounds) by the model

$$S = 256.2 - 47.9 \log w.$$

Approximate the number of strides per minute for a 500 pound horse when it breaks from a trot to a gallop.

29. ***Tornadoes*** The wind speed S (in miles per hour) near the center of a tornado is related to the distance d (in miles) the tornado travels by the model

$$S = 93 \log d + 65.$$

Approximate the wind speed of a tornado that traveled 150 miles.

Lesson 8.4

LESSON 8.4

NAME ______________________ DATE __________

Practice C

For use with pages 486–492

Rewrite the equation in exponential form.

1. $\log_5 125 = 3$
2. $\log_8 2 = \frac{1}{3}$
3. $\log_3 \frac{1}{27} = -3$

Use a calculator to evaluate the expression. Round the result to three decimal places.

4. $\ln 3 + 1$
5. $\dfrac{\ln 2.5}{10}$
6. $\dfrac{\log 4 - 3}{2}$

Evaluate the expression without using a calculator.

7. $\log_2 \frac{1}{32}$
8. $\log \frac{1}{1000}$
9. $\log_8 4$
10. $\log_{16} 8$
11. $\log_{27} \frac{1}{9}$
12. $\log_{100} \frac{1}{1000}$

Find the inverse of the function.

13. $f(x) = \log_4 x$
14. $f(x) = \log_2 (7x)$
15. $f(x) = \log (3x + 2)$
16. $f(x) = \ln x - 3$
17. $f(x) = \ln (x - 2) + 1$
18. $f(x) = \log 100^{x+2}$

Graph the function.

19. $f(x) = \log_3 x$
20. $f(x) = \log_3 (x + 2)$
21. $f(x) = \log_3 x - 1$
22. $f(x) = \log_3 (x + 2) - 1$
23. $f(x) = -\log_3 (x + 2)$
24. $f(x) = -\log_3 (x + 2) - 1$

Critical Thinking **In Exercises 25–28, use the following information.**

By definition of a logarithm, the base b of a logarithmic function must be a positive number and $b \neq 1$.

25. Assuming that $b = 1$, the "logarithmic function' would be written $y = \log_1 x$. Complete the table of values for this "logarithmic function."

x							
y	-2	-1	$-\frac{1}{2}$	0	$\frac{1}{2}$	1	2

26. Use the data to sketch a graph.
27. Does the graph look like a typical logarithmic graph?
28. Is the relation a function?

400-Meter Relay **In Exercises 29–31, use the following information.**

The winning time (in seconds) in the women's 400-meter relay at the Olympic Games from 1928 to 1996 can be modeled by the function $f(t) = 67.99 - 5.82 \ln t$, where t is the number of years since 1900.

29. In 1988 the United States team won the 400-meter relay. What was its winning time?
30. Use a graphing calculator to graph the model.
31. Use the graph to approximate the winning time in the 2000 Olympic Games.

NAME ______________________ DATE ________

Reteaching with Practice

For use with pages 486–492

GOAL Evaluate logarithmic functions, and graph logarithmic functions

VOCABULARY

Let b and y be positive numbers, $b \neq 1$. The **logarithm of y with base b** is denoted by $\log_b y$ and is defined as follows: $\log_b y = x$ if and only if $b^x = y$. The expression $\log_b y$ is read as "log base b of y."

The logarithm with base 10 is called the **common logarithm,** denoted by $\log_{10}$ or simply by log.

The logarithm with base e is called the **natural logarithm,** denoted by $\log_e$ or more often by ln.

If b is a positive real number such that $b \neq 1$, then $\log_b 1 = 0$ because $b^0 = 1$ and $\log_b b = 1$ because $b^1 = b$.

EXAMPLE 1 *Rewriting Logarithmic Equations*

Logarithmic Form	*Exponential Form*
a. $\log_{10} 1000 = 3$	$10^3 = 1000$
b. $\log_4 1 = 0$	$4^0 = 1$
c. $\log_9 \frac{1}{81} = -2$	$9^{-2} = \frac{1}{81}$

Exercises for Example 1

Rewrite the equation in exponential form.

1. $\log_4 64 = 4$ **2.** $\log_5 125 = 3$ **3.** $\log_7 1 = 0$

4. $\log_2 \frac{1}{8} = -3$ **5.** $\log_8 8 = 1$ **6.** $\log_{1/3} 3 = -1$

EXAMPLE 2 *Evaluating Logarithmic Expressions*

Evaluate the expressions (a) $\log_{27} 3$ and (b) $\log_6 216$.

SOLUTION

To evaluate a logarithm, you are finding an exponent. To help you evaluate $\log_b y$, ask yourself what power of b gives you y.

a. 27 to what power gives 3?

$27^{1/3} = 3$, so $\log_{27} 3 = \frac{1}{3}$.

b. 6 to what power gives 216?

$6^3 = 216$, so $\log_6 216 = 3$.

Exercises for Example 2

Evaluate the expression without using a calculator.

7. $\log_3 243$ **8.** $\log_2 2$ **9.** $\log_5 1$

10. $\log_{16} 4$ **11.** $\log_{1/3} 9$ **12.** $\log_{1/2} \frac{1}{32}$

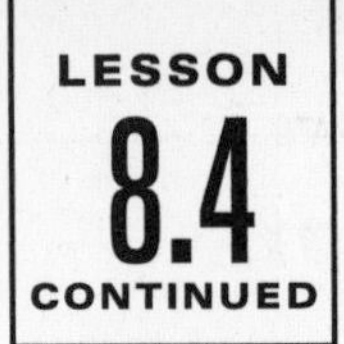

NAME ______________________ DATE ________

Reteaching with Practice

For use with pages 486–492

EXAMPLE 3 Using Inverse Properties

Simplify the expressions (a) $5^{\log_5 4}$ and (b) $\log_2 8^x$.

SOLUTION

a. $5^{\log_5 4} = 4$ — Use the inverse property $b^{\log_b x} = x$.

b. $\log_2 8^x = \log_2 (2^3)^x$ — Rewrite 8 as a power of the base 2.

$= \log_2 2^{3x}$ — Use power rule of exponents.

$= 3x$ — Use the inverse property $\log_b b^x = x$.

Exercises for Example 3

Simplify the expression.

13. $4^{\log_4 x}$ **14.** $8^{\log_8 10}$ **15.** $\log_6 6^x$ **16.** $\log_3 81^x$

EXAMPLE 4 Graphing Logarithmic Functions

Graph the function. State the domain and range.

a. $y = \log_3 x + 1$ **b.** $y = \ln (x - 2)$

SOLUTION

a. Because $h = 0$, the vertical line $x = 0$ is an asymptote. Plot the points (1, 1) and (3, 2). Because $b > 1$, from left to right, draw a curve that starts just to the right of the line $x = 0$ and moves up.

y
1
(1, 1)
(3, 2)
1
x
$y = \log_3 x + 1$

The domain is $x > 0$, and the range is all real numbers.

b. Because $h = 2$, the vertical line $x = 2$ is an asymptote. Plot the points (3, 0) and (5, 1.10). Because $b > 1$, from left to right, draw a curve that starts just to the right of the line $x = 2$ and moves up.

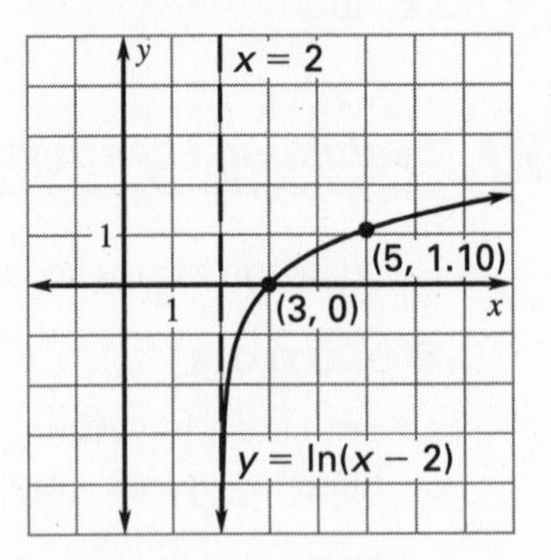

The domain is $x > 2$, and the range is all real numbers.

Exercises for Example 4

Graph the function. State the domain and range.

17. $y = \log_2 x$ **18.** $y = \log_{1/2} x$ **19.** $\ln (x + 2)$ **20.** $\ln x - 3$

Lesson 8.4

LESSON 8.4

NAME ______________________________ DATE __________

Quick Catch-Up for Absent Students

For use with pages 486–492

The items checked below were covered in class on (date missed) __________

Lesson 8.4: Logarithmic Functions

____ **Goal 1:** Evaluate logarithmic functions. (pp. 486, 487)

Material Covered:

____ Example 1: Rewriting Logarithmic Equations

____ Example 2: Evaluating Logarithmic Expressions

____ Example 3: Evaluating Common and Natural Logarithms

____ Example 4: Evaluating a Logarithmic Function

Vocabulary:

logarithm of y with base b, p. 486

common logarithm, p. 487

natural logarithm, p. 487

____ **Goal 2:** Graph logarithmic functions. (p. 488, 489)

Material Covered:

____ Student Help: Look Back

____ Example 5: Using Inverse Properties

____ Example 6: Finding Inverses

____ Example 7: Graphing Logarithmic Functions

____ Example 8: Using the Graph of a Logarithmic Function

____ Other (specify) ______________________________

____

Homework and Additional Learning Support

____ Textbook (specify) pp. 490–492

____

____ Internet: Extra Examples at www.mcdougallittell.com

____ *Reteaching with Practice* worksheet (specify exercises) __________

____ *Personal Student Tutor* for Lesson 8.4

Lesson 8.4

NAME ______________________ DATE __________

Cooperative Learning Activity

For use with pages 486–492

GOAL **To use a model of a logarithmic function to solve problems involving tornadoes**

Materials: calculator

Background

Mathematical models can be used to solve various problems. In these problems you will find either the speed of a tornado or the distance it traveled. Keep in mind that most tornadoes last less than an hour and travel less than 20 miles.

Instructions

1. Use the model

   $s = 93 \log d + 65,$

   where s equals the speed of the wind in miles per hour, and d represents the distance the tornado traveled in miles.

2. Find the speed of a tornado that traveled a distance of 220 miles.

3. Find the distance a tornado traveled if the speed was 250 miles per hour.

4. Find the speed of a tornado that traveled a distance of 150 miles.

5. Find the distance a tornado traveled if the speed was 175 miles per hour.

Analyzing the Results

1. How did your use of algebra enable you to solve these problems?

2. How was your calculator used?

NAME ______________________ DATE __________

Interdisciplinary Application

For use with pages 486–492

Heart Disease

HEALTH Coronary artery disease is the most common form of heart disease. The arteries that cover the surface of the heart carry oxygen to the heart muscle and when these arteries become clogged with fatty deposits, a condition called atherosclerosis, the heart is deprived of the oxygen and nutrients that it needs to work. A complete blockage of one of these arteries can result in a heart attack. In some cases, bypass surgery can be performed on a patient to attach new blood vessel segments to the coronary arteries that are blocked.

One of the many factors for heart disease is high blood cholesterol. A person's risk for coronary artery disease is a logarithmic function of that person's risk factor x, which can be calculated using the ratio

$$x = \frac{\text{total cholesterol}}{\text{HDL (High Density Lipoprotein) cholesterol}}.$$

The risk for men M and women W can then be modeled by

$M = 1.357 \ln x - 1.1875$ and

$W = 2.069 \ln x - 2.042.$

The risk calculated from these formulas is expressed as a multiple of "average" risk. For example, a person with a risk equal to 0.2 means that he or she has 20% of the average risk for coronary artery disease. The normal or average risk for Americans is a 75% chance of a heart attack in one's lifetime.

1. Use a graphing calculator to graph the models for M and W.
2. Find each patient's risk for coronary artery disease. Round your results to two decimal places.

   Patient A: Male with a total cholesterol of 180 and HDL cholesterol of 60

   Patient B: Female with a total cholesterol of 184 and HDL cholesterol of 46

   Patient C: Male with a total cholesterol of 210 and HDL cholesterol of 38

   Patient D: Female with a total cholesterol of 170 and HDL cholesterol of 52

   Patient E: Male with a total cholesterol of 196 and HDL cholesterol of 48

   Patient F: Female with a total cholesterol of 192 and HDL cholesterol of 42
3. Which patient from Exercise 2 has the highest risk of coronary artery disease? Which patient has the lowest?

NAME ______ DATE ______

Challenge: Skills and Applications

For use with pages 486–492

In Exercises 1–4, solve each equation.

1. $\log_9 x = \frac{3}{2}$

2. $\log_x \frac{1}{4} = -\frac{2}{3}$

3. $\log_{1/125} 3125 = x$

4. $\log_{1/16} x = \frac{5}{4}$

5. If k is a large positive integer, then

$$e \approx \left(1 + \frac{1}{k}\right)^k, \text{ and so } e^x \approx \left(1 + \frac{1}{k}\right)^{kx}.$$

a. Let x be any real number. Using the substitution $k = \frac{n}{x}$, show that

$$e^x \approx \left(1 + \frac{x}{n}\right)^n.$$

b. Substitute $\ln x$ for x in this formula and solve for $\ln x$.

c. Use the formula you found in part (b), with $n = 64$, to approximate $\ln 2$.

6. a. Find numerical values of log 7, log 70, log 700, and log 7000. Use these to predict the value of log 700,000.

b. Compare the numerical value of log 70 with that of $\log \sqrt{70}$. What do you notice?

c. Find numerical values of $\log \sqrt{7}$, $\log \sqrt{700}$, and $\log \sqrt{7000}$. Explain why $\log \sqrt{7}$ and $\log \sqrt{700}$ have the same decimal part, and $\log \sqrt{70}$ and $\log \sqrt{7000}$ have the same decimal part, but $\log \sqrt{7}$ and $\log \sqrt{70}$ do not.

7. Let b and x be positive numbers, with $b \neq 1$. Suppose that $\log_b x = y$. Use exponential form to prove that $\log_{1/b} x = -y$.

8. Suppose that $\log_b x = y$. Complete the following equation, in terms of n and y:

$\log_{\sqrt[n]{b}} x = ?$

Prove that the equation you wrote is true.

TEACHER'S NAME ______________________ CLASS ______ ROOM ______ DATE ______

Lesson Plan

2-day lesson (See *Pacing the Chapter,* TE pages 462C–462D) **For use with pages 493–500**

GOALS
1. **Use properties of logarithms.**
2. **Use properties of logarithms to solve real-life problems.**

State/Local Objectives ____

____

✓ Check the items you wish to use for this lesson.

STARTING OPTIONS
____ Homework Check: TE page 490; Answer Transparencies
____ Warm-Up or Daily Homework Quiz: TE pages 493 and 492, CRB page 67, or Transparencies

TEACHING OPTIONS
____ Motivating the Lesson: TE page 494
____ Lesson Opener (Activity): CRB page 68 or Transparencies
____ Graphing Calculator Activity with Keystrokes: CRB page 69
____ Examples: Day 1: 1–3, SE pages 493–494; Day 2: 4–5, SE pages 494–495
____ Extra Examples: Day 1: TE page 494 or Transp.; Day 2: TE pages 494–495 or Transp.
____ Technology Activity: SE page 500
____ Closure Question: TE page 495
____ Guided Practice: SE page 496 Day 1: Exs. 1–12; Day 2: Exs. 13

APPLY/HOMEWORK
Homework Assignment
____ Basic Day 1: 14–44 even, 46–51; Day 2: 52–57, 58–72 even, 74–76, 88–90, 92–102 even
____ Average Day 1: 14–21, 22–44 even, 46–55; Day 2: 56–84 even, 86–90, 92–102 even
____ Advanced Day 1: 14–21, 22–44 even, 46–57; Day 2: 58–84 even, 86–91, 92–102 even

Reteaching the Lesson
____ Practice Masters: CRB pages 70–72 (Level A, Level B, Level C)
____ Reteaching with Practice: CRB pages 73–74 or Practice Workbook with Examples
____ Personal Student Tutor

Extending the Lesson
____ Applications (Real-Life): CRB page 76
____ Math & History: SE page 499; CRB page 77; Internet
____ Challenge: SE page 498; CRB page 78 or Internet

ASSESSMENT OPTIONS
____ Checkpoint Exercises: Day 1: TE page 494 or Transp.; Day 2: TE pages 494–495 or Transp.
____ Daily Homework Quiz (8.5): TE page 499, CRB page 81, or Transparencies
____ Standardized Test Practice: SE page 498; TE page 499; STP Workbook; Transparencies

Notes ____

____

____

Lesson 8.5

TEACHER'S NAME ______________________ CLASS ________ ROOM ________ DATE ________

Lesson Plan for Block Scheduling

1-day lesson (See *Pacing the Chapter,* TE pages 462C–462D) **For use with pages 493–500**

GOALS

1. **Use properties of logarithms.**
2. **Use properties of logarithms to solve real-life problems.**

State/Local Objectives ______________________________

CHAPTER PACING GUIDE	
Day	**Lesson**
1	8.1 (all); 8.2(all)
2	8.3 (all)
3	8.4 (all)
4	**8.5 (all)**
5	8.6 (all)
6	8.7 (all); 8.8(all)
7	Review/Assess Ch. 8

✓ Check the items you wish to use for this lesson.

STARTING OPTIONS

____ Homework Check: TE page 490; Answer Transparencies
____ Warm-Up or Daily Homework Quiz: TE pages 493 and 492, CRB page 67, or Transparencies

TEACHING OPTIONS

____ Motivating the Lesson: TE page 494
____ Lesson Opener (Activity): CRB page 68 or Transparencies
____ Graphing Calculator Activity with Keystrokes: CRB page 69
____ Examples: 1–5: SE pages 493–495
____ Extra Examples: TE pages 494–495 or Transparencies
____ Technology Activity: SE page 500
____ Closure Question: TE page 495
____ Guided Practice Exercises: SE page 496

APPLY/HOMEWORK

Homework Assignment

____ Block Schedule: 14–21, 22–44 even, 46–55, 56–84 even, 86–90, 92–102 even

Reteaching the Lesson

____ Practice Masters: CRB pages 70–72 (Level A, Level B, Level C)
____ Reteaching with Practice: CRB pages 73–74 or Practice Workbook with Examples
____ Personal Student Tutor

Extending the Lesson

____ Applications (Real Life): CRB page 76
____ Math & History: SE page 499; CRB page 77; Internet
____ Challenge: SE page 498; CRB page 78 or Internet

ASSESSMENT OPTIONS

____ Checkpoint Exercises: TE pages 494–495 or Transparencies
____ Daily Homework Quiz (8.5): TE page 499, CRB page 81, or Transparencies
____ Standardized Test Practice: SE page 498; TE page 499; STP Workbook; Transparencies

Notes ______________________________

NAME ______________________ DATE ________

Available as a transparency

WARM-UP EXERCISES

For use before Lesson 8.5, pages 493–500

Simplify.

1. $\log 100 + \log 10{,}000$
2. $\log_5 25 + \log_5 125$
3. $\log_5 125 - \log_5 25$
4. $\log_5 25^2$
5. $2 \cdot \log_5 25$

DAILY HOMEWORK QUIZ

For use after Lesson 8.4, pages 486–492

1. Rewrite $\log_7 2401 = 4$ in exponential form.
2. Evaluate $\log_3 729$ without using a calculator.
3. Simplify $\log_5 625^x$.
4. Using a calculator, evaluate ln 7.4. Round to three deciamal places.
5. Find the inverse of $y = \ln(x - 0.5)$.
6. Graph $y = \log_3 x + 3$. State the domain and range.

LESSON 8.5

NAME ______________________ DATE __________

Activity Lesson Opener

For use with pages 493–500

Available as a transparency

SET UP: Work individually or in small groups

YOU WILL NEED: • scissors • a copy of this worksheet

Make a slide rule by cutting the figure below out from a copy of this page. Cut lengthwise along the center line so that you have two strips.

The slide rule uses a logarithmic scale. Each mark x is ($\ln x$) inches from the left edge of the slide rule. For example, mark 7 is ($\ln 7$) $\approx$ 1.95 inches from the left edge.

To solve $\ln 3 + \ln 4 = \ln x$, align mark 1 on the top strip with mark 3 on the bottom strip as shown. Find mark 4 on the top strip. It aligns with mark 12 on the bottom strip, so $\ln 3 + \ln 4 = \ln 12$.

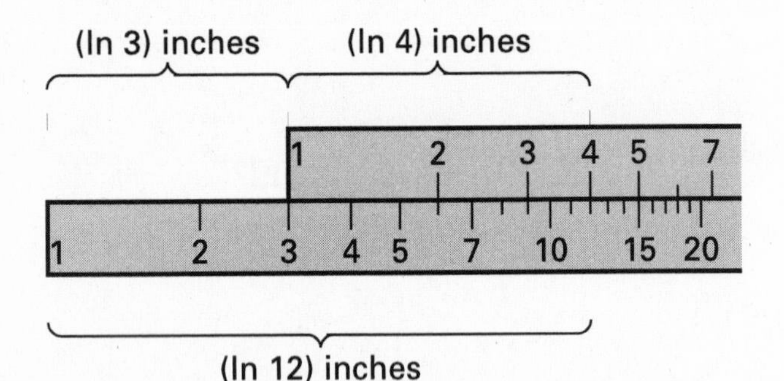

Use your slide rule to find the missing numbers.

1. $\ln 2 + \ln 3 = \ln$ ____
2. $\ln 2 + \ln 4 = \ln$ ____
3. $\ln 3 + \ln 5 = \ln$ ____
4. $\ln 3 + \ln 10 = \ln$ ____
5. $\ln 4 + \ln 5 = \ln$ ____
6. $\ln 4 + \ln 25 = \ln$ ____
7. $\ln 5 + \ln 5 = \ln$ ____
8. $\ln 7 + \ln 10 = \ln$ ____
9. $\ln a + \ln b = \ln$ ____ (*Hint:* Look for a pattern.)

LESSON 8.5

NAME ______________________ DATE __________

Graphing Calculator Activity Keystrokes

For use with page 500

TI-82

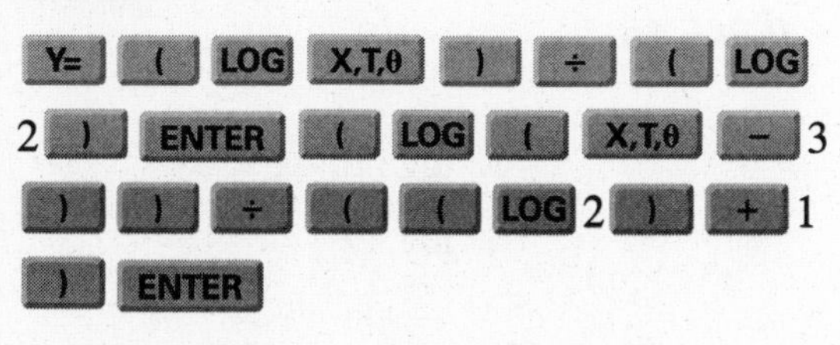
Y= (LOG X,T,θ) ÷ (LOG
2) ENTER (LOG (X,T,θ − 3
)) ÷ ((LOG 2) + 1
) ENTER

WINDOW ENTER (-) 1 ENTER 5
ENTER 1 ENTER (-) 8 ENTER 5 ENTER 1
ENTER GRAPH

TI-83

Y= (LOG X,T,θ,n) ÷ (LOG
2) ENTER
(LOG (X,T,θ,n − 3)) ÷
((LOG 2) + 1) ENTER

WINDOW (-) 1 ENTER 5 ENTER 1 ENTER
(-) 8 ENTER 5 ENTER 1 ENTER 1
ENTER GRAPH

SHARP EL-9600c

Y= (log X/θ/T/n) ÷ (log
2) ENTER
(log (X/θ/T/n − 3)) ÷
((log 2) + 1) ENTER

WINDOW (-) 1 ENTER 5
ENTER 1 ENTER (-) 8 ENTER 5 ENTER 1
ENTER GRAPH

CASIO CFX-9850GA PLUS

From the main menu, choose GRAPH.

(log X,θ,T) ÷ (log
2) EXE
(log (X,θ,T − 3)) ÷
((log 2) + 1) EXE

SHIFT F3 (-) 1 EXE 5 EXE 1 EXE (-) 8
EXE 5 EXE 1 EXE EXIT F6

LESSON 8.5

NAME ______________________ DATE ________

Practice A

For use with pages 493–499

Use the properties of logarithms to rewrite the expression in terms of log 2 and log 7. Then use log 2 ≈ 0.301 and log 7 ≈ 0.845 to approximate the expression.

1. $\log 4$
2. $\log 14$
3. $\log \left(\frac{7}{2}\right)$
4. $\log \left(\frac{2}{7}\right)$
5. $\log 7^{-3}$
6. $\log 49$

Expand the expression.

7. $\log_2(3x)$
8. $\log_3(9x)$
9. $\log\left(\frac{x}{5}\right)$
10. $\log_6\left(\frac{6}{x}\right)$
11. $\log_3 x^5$
12. $\ln x^{-3}$
13. $\log \sqrt[3]{x}$
14. $\log_2 \sqrt{2x}$
15. $\log_3(27x)^2$

Condense the expression.

16. $\log 3 + \log 5$
17. $\log_2 x + \log_2 7$
18. $\log_3 14 + \log_3 y$
19. $\log 4 - \log x$
20. $\ln x - \ln 3$
21. $\log (x - 1) - \log 6$
22. $\ln 2 - \ln (x + 2)$
23. $\log_3 (x + 5) + \log_3 4$
24. $2 \log x + \log 8$

Use the change-of-base formula to rewrite the expression. Then use a calculator to evaluate the expression. Round your result to three decimal places.

25. $\log_2 5$
26. $\log_7 10$
27. $\log_3 17$
28. $\log_6 200$
29. $\log_5 \frac{1}{2}$
30. $\log_4 1235$

Investments **In Exercises 31 and 32, use the following information.**

You want to invest in a stock whose value has been increasing by approximately 5% each year. The time required for an initial investment of I_0 to grow to I can be modeled by

$$t = \frac{\ln\left(\frac{I}{I_0}\right)}{0.049},$$

where I_0 and I are measured in dollars and t is measured in years.

31. Expand the expression for t.

32. Assume that you have \$1000 to invest. Complete the table to show how long your investment would take to double, triple, and quadruple.

I	2000	3000	4000
t			

NAME ______________________ DATE ________

Practice B

For use with pages 493–499

Use the properties of logarithms to rewrite the expression in terms of log 3 and log 4. Then use log 3 ≈ 0.477 and log 4 ≈ 0.602 to approximate the expression.

1. $\log\left(\frac{3}{4}\right)$

2. $\log 12$

3. $\log 9$

4. $\log 16$

5. $\log \frac{1}{4}$

6. $\log\left(\frac{4}{27}\right)$

Expand the expression.

7. $\log_6 3x$

8. $\log_2 \frac{x}{5}$

9. $\log xy^2$

10. $\log_4 \frac{xy}{3}$

11. $\log_3 \sqrt{x}\, y\, z$

12. $\log_5 2\sqrt{x}$

13. $\log \frac{x^2}{4}$

14. $\log \frac{10}{\sqrt{x}}$

15. $\log_2 \frac{x^2y}{z}$

Condense the expression.

16. $\log_3 7 - \log_3 x$

17. $2\log_5 x + \log_5 3$

18. $\log_4 5 + \log_4 x + \log_4 y$

19. $\frac{1}{2}\log x - \log 4$

20. $\frac{2}{3}\log_2 x - 3\log_2 y$

21. $\log_3 4 + 2\log_3 x - \log_3 5$

Use the change-of-base formula to rewrite the expression. Then use a calculator to evaluate the expression. Round your result to three decimal places if necessary.

22. $\log_3 12$

23. $\log_6 2$

24. $\log_4 0.5$

25. $\log_{0.8} 12$

26. $\log_{1.5} 2.8$

27. $\log_{1/2} 6$

Henderson-Hasselbach Formula **In Exercises 28–32, use the following information.**

The pH of a patient's blood can be calculated using the Henderson-Hasselbach Formula, $pH = 6.1 + \log \frac{B}{C}$, where B is the concentration of bicarbonate and C is the concentration of carbonic acid. The normal pH of blood is approximately 7.4.

28. Expand the right side of the formula.

29. A patient has a bicarbonate concentration of 24 and a carbonic acid concentration of 1.9. Find the pH of the patient's blood.

30. Is the patient's pH in Exercise 29 below normal or above normal?

31. A patient has a bicarbonate concentration of 24. Graph the model.

32. Use the graph to approximate the concentration of carbonic acid required for the patient to have normal blood pH.

NAME ______________________________ DATE __________

Practice C

For use with pages 493–499

Use the properties of logarithms to rewrite the expression in terms of ln 2, ln 3, and ln 5. Then use ln 2 ≈ 0.693, ln 3 ≈ 1.099, and ln 5 ≈ 1.609 to approximate the expression.

1. $\ln 6$

2. $\ln\left(\frac{10}{3}\right)$

3. $\ln 30$

4. $\ln 12$

5. $\ln\left(\frac{2}{5}\right)$

6. $\ln\left(\frac{5}{6}\right)$

Expand the expression.

7. $\log(8x)$

8. $\log_3 xyz$

9. $\log_4 \frac{2xy}{z}$

10. $\ln \frac{x}{yz}$

11. $\log \sqrt{3xy}$

12. $\log_5 \frac{\sqrt{x}}{y}$

13. $\ln \frac{3y}{\sqrt[4]{x}}$

14. $\log(3xyz^2)^3$

15. $\log_2 \frac{(xy)^4}{z^2}$

Condense the expression.

16. $\log 3 - \log 4 - \log 7$

17. $\ln x - \ln y + \ln z + \ln 3$

18. $3 \ln x - 2 \ln y - 4 \ln z$

19. $\log_2(x - 4) + 5 \log_2(x + 1) - 3 \log_2(x - 1)$

20. $\frac{1}{2}\log(x + 5) - 2 \log x + \ln y$

21. $3[\ln(x - 2) + 2 \ln(x + 1) - \ln(x + 2) - 5 \ln(x - 1)]$

Use the change-of-base formula to rewrite the function in terms of common (base 10) or natural (base ln) logarithms.

22. $y = \log_3 x$

23. $y = \log_6(x + 3)$

24. $y = \log_2(x - 1) + 3$

Annuities **In Exercises 25 and 26, use the following information.**

An ordinary annuity is an account in which you make a fixed deposit at the end of each compounding period. You want to use an annuity to help you save money for college. The formula

$$t = \frac{\ln\left[\frac{Sr + Pn}{Pn}\right]}{n \ln\left[\frac{n + r}{n}\right]}$$

gives the time t (in years) required to have S dollars in the annuity if your periodic payments P (in dollars) are made n times a year and the annual interest rate is r (in decimal form).

25. Expand the right side of the formula.

26. How long will it take you to save $20,000 in annuity that earns an annual interest rate of 5% if you make monthly payments of $50?

NAME ______________________________ DATE __________

Reteaching with Practice

For use with pages 493–499

GOAL **Use properties of logarithms**

VOCABULARY

Properties of Logarithms
Let b, u, and v be positive numbers such that $b \neq 1$.

Product Property $\log_b uv = \log_b u + \log_b v$

Quotient Property $\log_b \frac{u}{v} = \log_b u - \log_b v$

Power Property $\log_b u^n = n \log_b u$

Change-of-Base Formula Let u, b, and c be positive numbers with $b \neq 1$ and $c \neq 1$. Then: $\log_c u = \frac{\log_b u}{\log_b c}$.

In particular, $\log_c u = \frac{\log u}{\log c}$ and $\log_c u = \frac{\ln u}{\ln c}$.

EXAMPLE 1 *Using Properties of Logarithms*

Use $\log_3 2 \approx 0.631$ and $\log_3 5 \approx 1.465$ to approximate the following.

a. $\log_3 \frac{2}{5}$ **b.** $\log_3 10$ **c.** $\log_3 125$

SOLUTION

a. $\log_3 \frac{2}{5} = \log_3 2 - \log_3 5 \approx 0.631 - 1.465 = -0.834$

b. $\log_3 10 = \log_3 (2 \cdot 5) = \log_3 2 + \log_3 5 \approx 0.631 + 1.465 = 2.096$

c. $\log_3 125 = \log_3 5^3 = 3 \log_3 5 \approx 3(1.465) = 4.395$

Exercises for Example 1

Use $\log_6 4 \approx 0.774$ and $\log_6 10 \approx 1.285$ to approximate the value of the expression.

1. $\log_6 40$ **2.** $\log_6 100$ **3.** $\log_6 \frac{10}{4}$ **4.** $\log_6 64$

EXAMPLE 2 *Expanding a Logarithmic Expression*

Expand $\ln 6x^5$. Assume x is positive.

SOLUTION

$\ln 6x^5 = \ln 6 + \ln x^5$ Product Property

$= \ln 6 + 5 \ln x$ Power Property

NAME ______________________ DATE ________

Reteaching with Practice

For use with pages 493–499

Exercises for Example 2

Expand the expression.

5. $\log 9x$ **6.** $\log_2 6x^3$ **7.** $\log_6 \frac{2}{3}$

8. $\log_3 \frac{4x}{5}$ **9.** $\ln 2xy$ **10.** $\ln \frac{2x^2}{y}$

EXAMPLE 3 Condensing a Logarithmic Expression

Condense $3 \ln x + \ln 4 - \ln 7x$.

SOLUTION

$$3 \ln x + \ln 4 - \ln 7x = \ln x^3 + \ln 4 - \ln 7x \quad \text{Power Property}$$

$$= \ln (x^3 \cdot 4) - \ln 7x \quad \text{Product Property}$$

$$= \ln \frac{4x^3}{7x} \quad \text{Quotient Property}$$

$$= \ln \frac{4x^2}{7} \quad \text{Simplify.}$$

Exercises for Example 3

Condense the expression.

11. $\log_4 12 + \log_4 5$ **12.** $\log x - \log y$ **13.** $\ln 3 + \ln 6 - \ln 9$

14. $3 \log_2 3$ **15.** $6 \log_2 x + 3 \log_2 x$ **16.** $\ln 24 - 3 \ln 2$

EXAMPLE 4 Using the Change-of-Base Formula

Evaluate the expression $\log_2 9$ using common and natural logarithms.

SOLUTION

Notice that the base of the logarithm is two. Most scientific calculators can only evaluate common logarithms of base ten and natural logarithms of base e. You must use the change-of-base formula.

Using common logarithms: $\log_2 9 = \frac{\log 9}{\log 2} \approx \frac{0.9542}{0.3010} \approx 3.170$

Using natural logarithms: $\log_2 9 = \frac{\ln 9}{\ln 2} \approx \frac{2.1972}{0.6931} \approx 3.170$

Notice that you obtain the same result using either common or natural logarithms.

Exercises for Example 4

Use the change-of-base formula to evaluate the expression.

17. $\log_3 30$ **18.** $\log_4 13$ **19.** $\log_2 17$ **20.** $\log_5 10$

NAME ______________________________ DATE __________

Quick Catch-Up for Absent Students

For use with pages 493–500

The items checked below were covered in class on (date missed) ____________

Lesson 8.5: Properties of Logarithms

____ **Goal 1:** Use properties of logarithms. (pp. 493, 494)

Material Covered:

____ Activity: Investigating a Property of Logarithms

____ Example 1: Using Properties of Logarithms

____ Example 2: Expanding a Logarithmic Expression

____ Example 3: Condensing a Logarithmic Expression

____ Example 4: Using the Change-of-Base Formula

____ **Goal 2:** Use properties of logarithms to solve real-life problems. (p. 495)

Material Covered:

____ Example 5: Using properties of Logarithms

Activity 8.5: Graphing Logarithmic Functions (p. 500)

____ **Goal 1:** Use a graphing calculator to graph logarithmic functions.

____ Student Help: Keystroke Help

____ Other (specify) ____

____

Homework and Additional Learning Support

____ Textbook (specify) pp. 496–499

____

____ *Reteaching with Practice* worksheet (specify exercises)______________

____ *Personal Student Tutor* for Lesson 8.5

NAME ______________________________ DATE __________

Real-Life Application: When Will I Ever Use This?

For use with pages 493–499

Home Mortgages

People who purchase homes usually make a down payment and finance the remainder of the cost of their home through a home mortgage. The amount financed for a home is usually paid by means of mortgaging the house through an installment loan with equal monthly payments over a period of 20 to 35 years. The amount of a homeowner's monthly payment depends on the amount that was financed, the length of the mortgage, and the annual percent rate. People who are interested in purchasing a home may become concerned about what their monthly payment might be, so in order to keep their monthly payment low, they are willing to take out a loan over a very long period of time such as 30 or 35 years.

The amount of interest paid in a typical home mortgage can be an enormous amount. For example, if $100,000 were financed for 30 years at an annual percent rate of 9%, the homeowner would end up paying more than $280,000 by the time the mortgage was paid off!

In Exercises 1–5, use the following information.

The length t (in years) of a home mortgage of $150,000 at an annual percent rate of 8% can be modeled by

$$t = 12.542 \ln\left(\frac{x}{x - 1000}\right), \quad x > 1000,$$

where x is the monthly payment in dollars.

1. Graph the model. What is the vertical asymptote of the graph?
2. Expand the expression for t.
3. Approximate the length of a $150,000 mortgage at 8% if the monthly payment is $1157.72.
4. Approximate the length of a $150,000 mortgage at 8% if the monthly payment is $1100.65.
5. Approximate the total amount paid over the term of the mortgage with a monthly payment of $1157.72. How much of the amount paid is interest?

NAME ______________________________ DATE __________

Math and History Application

For use with page 499

HISTORY The Math and History feature on page 499 mentions that logarithms are used to measure energies that span many orders of magnitude, like sound energy and energy released by earthquakes. Logarithms are also used to measure something important that you can't hear or feel: information. The branch of mathematics called information theory, founded in 1948 by Claude Shannon of Bell Labs, uses base 2 logarithms to measure the information content of a message.

The units of information are called bits. Every time you get an answer to a yes/no question, you have acquired one bit of information. Here's an example based on a dictionary game you might have played.

> A friend picks a word from the 20,000 words in a dictionary, and asks you to guess the word in as few guesses as possible. Of course you might have an idea of your friend's favorite word. But if you have no idea what word your friend might have picked, the best strategy is to start with a word right in the middle of the dictionary and ask "Does your word come before this word in the dictionary?" Whatever your friend answers, you now know which half of the dictionary holds your word. Now you can cut this in half again with another yes/no question, and keep going until you have zeroed in on the word.

How many guesses should it take you? The answer is surprisingly small: $\log_2 20{,}000$, which is about 14.3. Rounding up, you can see that if you follow this strategy you will *always* be able to guess your friend's word in 15 guesses. The *information content* of your friend's choice is 15 bits.

MATH Here are some problems about using logarithms to measure information.

1. If someone asks you to guess a number between 1 and a million and you use the "cut in half" strategy, what will your first question be? How many guesses will it take you? [*Hint:* The number of guesses required may remind you of another famous guessing game.]

2. Computer storage and memory is measured in 8-bit chunks called bytes. Each computer bit is a switch that is either on or off. When you see a memory dump from a computer, the bits are represented by 0s and 1s. You can think of each bit as being the answer to a yes/no question, with 0 standing for no and 1 for yes. How many bits does it take to code one character from a standard computer keyboard which has 47 alphanumeric keys, each of which can be either unshifted (like lowercase u or the number 4) or shifted (like U or the dollar sign $)? Is this more or less than one byte?

3. How many bits does it take to identify uniquely every person in the United States (the current population is about 300 million)?

NAME ______________________ DATE __________

Challenge: Skills and Applications

For use with pages 493–499

In Exercises 1–6, suppose $\log_b 3 = p$ and $\log_b 5 = q$. Express each quantity in terms of p and q.

1. $\log_b 25$
2. $\log_b \frac{1}{15}$
3. $\log_b \sqrt{3}$
4. $\log_b 45$
5. $\log_b \frac{5}{b}$
6. $\log_b 3b^2$
7. Suppose you know that $\log 2 = 0.301$ and you want to construct a table of approximate logarithms of the integers between 1 and 9.
   a. Find log 4, log 5, and log 8 directly from the value of log 2.
   b. Use the fact that $3^4 \approx 80$ to approximate log 3, log 6, and log 9.
   c. Use the fact that $3 \cdot 7^3 \approx 2^{10}$ to approximate log 7.
8. In this exercise you will prove the change-of-base formula. Let u, b, and c be positive numbers with $b \neq 1$ and $c \neq 1$.
   a. Let $x = \log_b u$ and $y = \log_b c$. Write these equations in exponential form.
   b. Solve the second equation you wrote in part (a) for b, and substitute this expression for b in the first equation, to prove the change-of-base formula.
9. Use the change-of-base formula to prove the following fact:

   $$\log_b a = \frac{1}{\log_a b}.$$

10. Nicholas Mercator (1620–1687) used the number $g = 1.005$, which is close to 1, as a base for a system of logarithms. By multiplication, he found that $g^{139} \approx 2$.
    a. Use this fact to find the values of $\log_g 0.5$, $\log_g 8$, and $\log_g 64$.
    b. Explain how Mercator could use his system of logs to calculate the product $64 \cdot 128$, without actually multiplying the two numbers.

LESSON 8.6

TEACHER'S NAME ______________ CLASS ______ ROOM ______ DATE ______

Lesson Plan

2-day lesson (See *Pacing the Chapter,* TE pages 462C–462D) **For use with pages 501–508**

GOALS

1. **Solve exponential equations.**
2. **Solve logarithmic equations.**

State/Local Objectives ______________________________

____

✓ Check the items you wish to use for this lesson.

STARTING OPTIONS

____ Homework Check: TE page 496; Answer Transparencies
____ Warm-Up or Daily Homework Quiz: TE pages 501 and 499, CRB page 81, or Transparencies

TEACHING OPTIONS

____ Motivating the Lesson: TE page 502
____ Lesson Opener (Graphing Calculator): CRB page 82 or Transparencies
____ Examples: Day 1: 1–4, SE pages 501–502; Day 2: 5–8, SE pages 503–504
____ Extra Examples: Day 1: TE page 502 or Transp.; Day 2: TE pages 503–504 or Transp.; Internet
____ Closure Question: TE page 504
____ Guided Practice: SE page 505 Day 1: Exs. 4–9, 16; Day 2: Exs. 1–3, 10–15, 17–18

APPLY/HOMEWORK

Homework Assignment

____ Basic Day 1: 23–33, 34–42 even, 62, 63, 65; Day 2: 19–22, 44–60 even, 61, 64, 69, 71, 77–87 odd; Quiz 2: 1–17
____ Average Day 1: 23–40, 62–68 even; Day 2: 19–22, 44–60 even, 61, 67–71 odd, 77–87 odd; Quiz 2: 1–17
____ Advanced Day 1: 23–42, 62–68; Day 2: 19–22, 44–60 even, 61, 67–71 odd, 72–76, 77–87 odd; Quiz 2: 1–17

Reteaching the Lesson

____ Practice Masters: CRB pages 83–85 (Level A, Level B, Level C)
____ Reteaching with Practice: CRB pages 86–87 or Practice Workbook with Examples
____ Personal Student Tutor

Extending the Lesson

____ Applications (Interdisciplinary): CRB page 89
____ Challenge: SE page 507; CRB page 90 or Internet

ASSESSMENT OPTIONS

____ Checkpoint Exercises: Day 1: TE page 502 or Transp.; Day 2: TE pages 503–504 or Transp.
____ Daily Homework Quiz (8.6): TE page 508, CRB page 94, or Transparencies
____ Standardized Test Practice: SE page 507; TE page 508; STP Workbook; Transparencies
____ Quiz (8.4–8.6): SE page 508; CRB page 91

Notes ______________________________

____

____

TEACHER'S NAME ______________________ CLASS ______ ROOM ______ DATE ______

Lesson Plan for Block Scheduling

1-day lesson (See *Pacing the Chapter,* TE pages 462C–462D) **For use with pages 501–508**

1. Solve exponential equations.
2. Solve logarithmic equations.

State/Local Objectives ______________________

CHAPTER PACING GUIDE	
Day	**Lesson**
1	8.1 (all); 8.2(all)
2	8.3 (all)
3	8.4 (all)
4	8.5 (all)
5	**8.6 (all)**
6	8.7 (all); 8.8(all)
7	Review/Assess Ch. 8

✓ Check the items you wish to use for this lesson.

STARTING OPTIONS

____ Homework Check: TE page 496; Answer Transparencies
____ Warm-Up or Daily Homework Quiz: TE pages 501 and 499, CRB page 81, or Transparencies

TEACHING OPTIONS

____ Motivating the Lesson: TE page 502
____ Lesson Opener (Graphing Calculator): CRB page 82 or Transparencies
____ Examples: 2–8: SE pages 501–504
____ Extra Examples: TE pages 502–504 or Transarencies; Internet
____ Closure Question: TE page 504
____ Guided Practice Exercises: SE page 505

APPLY/HOMEWORK

Homework Assignment

____ Block Schedule: 19–24, 26–60 even, 61–71 odd, 77–87 odd; Quiz 2: 1–17

Reteaching the Lesson

____ Practice Masters: CRB pages 83–85 (Level A, Level B, Level C)
____ Reteaching with Practice: CRB pages 86–87 or Practice Workbook with Examples
____ Personal Student Tutor

Extending the Lesson

____ Applications (Interdisciplinary): CRB page 89
____ Challenge: SE page 507; CRB page 90 or Internet

ASSESSMENT OPTIONS

____ Checkpoint Exercises: TE pages 502–504 or Transparencies
____ Daily Homework Quiz (8.6): TE page 508, CRB page 94, or Transparencies
____ Standardized Test Practice: SE page 507; TE page 508; STP Workbook; Transparencies
____ Quiz (8.4–8.6): SE page 508; CRB page 91

Notes ______________________

LESSON

8.6

NAME ______________________ DATE ________

Available as a transparency

Warm-Up Exercises

For use before Lesson 8.6, pages 501–508

1. Write 27^{4x} as a power of 3.

2. Evaluate $\log_5 4$.

3. Simplify $(4^2)^{2x-3}$.

Complete each statement.

4. If $3^x = 5$, then $\log_3 3^x =$ ____.

5. If $10^{\log(2x)} = 10^3$, then $\log(2x) =$ ____.

Lesson 8.6

Daily Homework Quiz

For use after Lesson 8.5, pages 493–500

1. Use a property of logarithms to evaluate $\log_2 8^4$.

2. Use $\log 5 \approx 0.699$ and $\log 6 \approx 0.778$ to approximate the value of $\log \frac{1}{150}$.

3. Expand $\ln 7^{1/4} 2x^{2/3}$.

4. Condense $4 \log_5 10 - 2 \log_5 50$.

5. Use the change-of-base formula to evaluate $\log_8 20$.

NAME ________________________ DATE __________

Graphing Calculator Lesson Opener

For use with pages 501–508

In Lesson 8.6, you will learn algebraic methods for solving equations that contain logarithms and exponents. You can use a graphing calculator to obtain appropriate solutions to these equations.

For example, to solve $5^{x-2} = 3 - 2^x$, graph $y_1 = 5^{x-2}$ and $y_2 = 3 - 2^x$. Use the *Intersect* feature. The solution is $x \approx 1.39$.

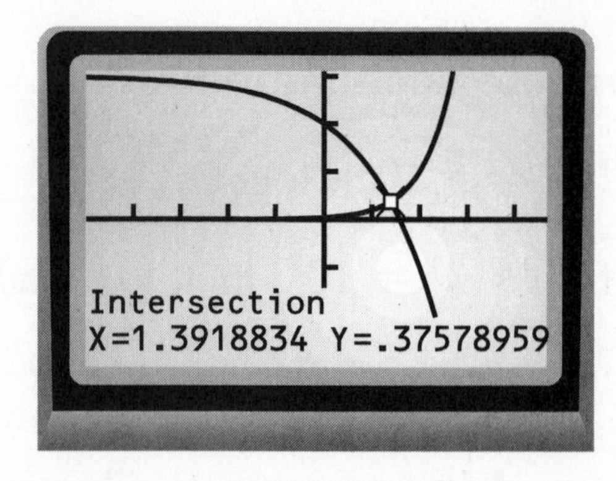

The number of solutions may vary. You may need to change the viewing window to find all solutions.

Use a graphing calculator to solve the equation. If necessary, round to the nearest hundredth.

1. $2^{x+2} = 10^x - 3$

2. $3^{1-x} = 4 - 2^x$

3. $5 \log (2x - 3) = 2$

4. $\ln (x + 3) + \ln (5 - x) = 1$

5. $3^{3x-2} = 2^x - 3$

6. $2^x - 5^{x-3} = 6$

7. $\ln (3x + 4) = 6$

8. $3 \log 2x = x - 5$

NAME ______________________ DATE ________

Practice A

For use with pages 501–508

Tell whether the *x*-value is a solution of the equation.

1. $\ln x = 9, x = e^9$
2. $\ln x = 3, x = 3e$
3. $\ln x = 7, x = 7^e$
4. $\ln 2x = 8, x = e^8$
5. $\ln 6x = 4, x = \frac{e^4}{6}$
6. $\ln 2x = 14, x = 2e^{14}$

Tell whether the *x*-value is a solution of the equation.

7. $e^x = 5, x = 5$
8. $e^x = 7, x = \ln 7$
9. $e^x = 3, x = \log 3$
10. $2e^x = 8, x = \ln 4$
11. $3e^x - 1 = 11, x = 4$
12. $5e^x + 2 = 17, x = \ln 3$

Solve the equation.

13. $4^x = 4^{2x+1}$
14. $3^{2x} = 3^{x-5}$
15. $2^{4x+1} = 2^{2x-3}$
16. $e^{3x} = e^{2x+7}$
17. $e^{2x-1} = e^{3-x}$
18. $10^x = 10^{7-3x}$

Solve the equation by taking the appropriate log of each side.

19. $2^x = 9$
20. $3^x = 10$
21. $e^x = 5$
22. $e^{2x} = 6$
23. $2^x + 5 = 12$
24. $5^{3x} - 2 = 8$

Use the following property to solve the equation. For positive numbers *b*, *x*, and *y* where $b \neq 1$, $\log_b x = \log_b y$ if and only if $x = y$.

25. $\log x = \log 7$
26. $\log(x + 2) = \log 9$
27. $\log_2(4x) = \log_2 12$
28. $\log_3(x - 1) = \log_3(2x + 5)$
29. $\ln(x + 3) = \ln(6 - 3x)$
30. $\log(3x + 2) = \log(x - 1)$

Solve the equation by exponentiating each side.

31. $\log_2 x = 5$
32. $\log_3(x - 1) = 8$
33. $\log(2x + 3) = 6$
34. $\ln(5x - 3) = 2$
35. $\ln(3x + 1) = 0$
36. $\log(4x) + 1 = 3$

Compound Interest **You deposit $100 in an account that earns 3% annual interest compounded continuously. How long does it take the balance to reach the following amounts?**

37. $110
38. $150
39. $200

NAME ______________________ DATE __________

Practice B

For use with pages 501–508

Solve the exponential equation. Round the result to three decimal places if necessary.

1. $e^x = 18$
2. $10^x = 350$
3. $e^{2x} = 42$
4. $e^x + 3 = 8$
5. $2^x + 7 = 10$
6. $5^{2x} = 8$
7. $2^{3x} = 4$
8. $e^{2x} = 5$
9. $3^{2x} - 3 = 4$
10. $e^{3x} + 6 = 10$
11. $e^{4x} - 3 = 7$
12. $2^{-x} + 1 = 6$
13. $4^{-2x} - 3 = 1$
14. $e^{-2x} + 5 = 12$
15. $e^{-x} - 6 = 1$
16. $2e^x = 10$
17. $4(2^x) = 16$
18. $3e^{-x} = 18$
19. $2e^{4x} = 5$
20. $3e^{5x} = 14$
21. $2(2^{3x}) = 2$
22. $-4e^{2x} + 3 = -5$
23. $-3e^{-x} - 4 = -13$
24. $2^{0.1x} + 6 = 12$
25. $\frac{1}{3}e^x + 1 = 5$
26. $\frac{2}{3}e^{2x} = 12$
27. $\frac{3}{8}(2^{3x}) + 1 = 10$

Solve the logarithmic equation. Round the result to three decimal places if necessary.

28. $\ln x = 5$
29. $\log_{10} x = -2$
30. $\log_2 x = 1.5$
31. $7 \ln x = 21$
32. $2 \log_{10} x = 10$
33. $7 + \log_{10} x = 4$
34. $-3 + \ln x = 5$
35. $4 - \ln x = 1$
36. $-5 + 2 \ln x = 5$
37. $3 \log_{10} x + 1 = 13$
38. $9 \log_{10} x - 4 = 11$
39. $\log_3 3x = 2$
40. $\log_2 5x = 1$
41. $2 + \log_3 2x = -3$
42. $\ln 4x - 6 = 8$
43. $2 + \log_2 3x = 8$
44. $\log_2 (x + 2) = \log_2 3x$
45. $\log_3 (2x + 1) = \log_3 (x - 4)$
46. $\ln (5x - 1) = \ln (3x + 2)$
47. $\ln (2x + 3) = \ln (2x - 1)$
48. $\ln (4x - 9) = \ln x$

49. ***Compound Interest*** You deposit \$2000 into an account that pays 2% annual interest compounded quarterly. How long will it take for the balance to reach \$2500?

50. ***Rocket Velocity*** Disregarding the force of gravity, the maximum velocity v of a rocket is given by $v = t \ln M$, where t is the velocity of the exhaust and M is the ratio of the mass of the rocket with fuel to its mass without fuel. A solid propellant rocket has an exhaust velocity of 2.5 kilometers per second. Its maximum velocity is 7.5 kilometers per second. Find its mass ratio M.

NAME ______________________ DATE __________

Practice C

For use with pages 501–508

Solve the exponential equation. Round the result to three decimal places if necessary.

1. $e^x = 9$
2. $2^{3x+1} = 4$
3. $3^{2x-5} = 7$
4. $e^{4x+1} - 3 = 8$
5. $e^{5-3x} + 4 = 6$
6. $3^{0.4x} - 7 = 10$
7. $\frac{2}{3}e^{4x} + 5 = 8$
8. $\frac{1}{4}(2^{3x+1}) - 2 = 5$
9. $\frac{5}{3}e^{1-x} + 1 = \frac{9}{2}$
10. $e^{x^2} + 3 = 4$
11. $e^{x^2+1} = e^{x+3}$
12. $2^{3x+1} = 2^{2/x}$

Solve the logarithmic equation. Round the result to three decimal places if necessary.

13. $\log(2x + 1) = 1$
14. $\ln(x + 3) - 2 = 8$
15. $\log_3(x - 2) + 5 = 7$
16. $\ln(6x + 5) = 7$
17. $\ln(x - 2) + \ln x = 0$
18. $\log_2 x + \log_2(x + 1) = 1$
19. $\log_3 x + \log_3(x - 2) = 1$
20. $\log_2(x + 1) - \log_2 x = 3$
21. $\log_4(x + 2) - \log_4(x - 3) = 2$
22. $\log(3x + 2) = \log(2x - 1)$
23. $\log(x^2 - 1) = \log(x + 5)$
24. $\log(x + 2) + \log(x - 3) = \log(x + 29)$
25. $\log_2 x + \log_2(x - 2) - \log_2(x - 3) = 3$
26. $\log_2(-x - 3) - \log_2(x - 1) - \log_2(x + 3) = 1$

Solve the exponential equation. Round the result to three decimal places.

27. $2^{x+1} = 3^{2x}$
28. $e^{x-3} = 10^{4-x}$
29. $5^{2x+1} = 2^{4x-3}$

Solve the logarithmic equation. Round the result to three decimal places.

30. $\log_2(x + 1) = \log_4(2x - 3)$
31. $\log_3(x - 3) = \log_9 x$
32. $\log(x - 4) = \log_{100}(x + 3)$

33. ***Compound Interest*** You deposit $2500 into an account that pays 3.5% annual interest compounded daily. How long will it take for the balance to reach $3000?

Loan Repayment **In Exercises 34–36, use the following information.**

The formula $L = P\left[\dfrac{1 - \left(1 + \dfrac{r}{n}\right)^{-nt}}{\dfrac{r}{n}}\right]$ gives the amount of a loan L in terms of the amount of each payment P, the interest rate r, the number of payments per year n, and the number of years t.

34. When purchasing a home, you need a loan for $80,000. The interest rate of the loan is 8% and you are required to make monthly payments of $587. How long will it take you to pay off the loan?
35. When the loan is paid off, how much money will you have paid the bank?
36. How much did you pay in interest?

NAME ______________________ DATE ____________

Reteaching with Practice

For use with pages 501–508

GOAL **Solve exponential equations and logarithmic equations**

> **VOCABULARY**
>
> For $b > 0$ and $b \neq 1$, if $b^x = b^y$, then $x = y$.
>
> For positive numbers b, x, and y where $b \neq 1$, $\log_b x = \log_b y$ if and only if $x = y$.

EXAMPLE 1 *Solving by Equating Exponents*

Solve $9^{x+1} = 27^{x-1}$.

SOLUTION

$9^{x+1} = 27^{x-1}$	Write original equation.
$(3^2)^{x+1} = (3^3)^{x-1}$	Rewrite each power with base 3.
$3^{2x+2} = 3^{3x-3}$	Power of a power property
$2x + 2 = 3x - 3$	Equate exponents
$x = 5$	Solve for x.

The solution is 5.

Exercises for Example 1

Solve the equation.

1. $5^{3x} = 5^{x+8}$

2. $10^{2x+3} = 10^{4x-1}$

3. $25^{2x+(1/2)} = 125^x$

4. $16 = 4^{x+1}$

EXAMPLE 2 *Taking a Logarithm of Each Side*

Solve $e^{-x} - 6 = 9$.

SOLUTION

Notice that you cannot rewrite each number with the same base. You can solve the equation by taking a logarithm of each side.

$e^{-x} - 6 = 9$	Write original equation.
$e^{-x} = 15$	Add 6 to each side.
$\ln e^{-x} = \ln 15$	Take natural log of each side.
$-x = \ln 15$	$\ln e^x = x$
$x \approx -2.708$	Divide each side by -1 and use a calculator.

The solution is about -2.708.

NAME ______________________ DATE __________

Reteaching with Practice

For use with pages 501–508

Exercises for Example 2

Solve the equation.

5. $5^x = 8$
6. $e^{-x} = 5$
7. $2^x + 1 = 5$
8. $10^{2x} - 6 = 146$
9. $9 - 4e^x = 5$
10. $\frac{1}{2}e^{-2x} = 6$

EXAMPLE 3 Solving a Logarithmic Equation

Solve $\ln (2x + 3) = \ln (5x - 6)$.

SOLUTION

$\ln (2x + 3) = \ln (5x - 6)$	Write original equation.
$2x + 3 = 5x - 6$	$\log_b x = \log_b y$ implies $x = y$.
$9 = 3x$	Subtract $2x$ and add 6 to each side.
$3 = x$	Divide each side by 3.

The solution is 3.

Exercises for Example 3

Solve the equation.

11. $\log (x + 3) = \log (3x + 1)$
12. $\log_2 (x - 1) = \log_2 (2x + 1)$
13. $\ln (4 - x) = \ln (4x - 11)$

EXAMPLE 4 Exponentiating Each Side

Solve $4 \log_3 3x = 20$.

SOLUTION

$4 \log_3 3x = 20$	Write original equation.
$\log_3 3x = 5$	Divide each side by 4.
$3^{\log_3 3x} = 3^5$	Exponentiate each side using base 3.
$3x = 243$	$b^{\log_b x} = x$
$x = 81$	Solve for x.

The solution is 81.

Exercises for Example 4

Solve the equation.

14. $\log_8 (x - 5) = \frac{2}{3}$
15. $3 \log_5 (x + 2) = 6$
16. $4 \ln 2x = 5$

NAME ______________________________ DATE ____________

Quick Catch-Up for Absent Students

For use with pages 501–508

Lesson 8.6

The items checked below were covered in class on (date missed) ____________

Lesson 8.6: Solving Exponential and Logarithmic Equations

____ **Goal 1:** Solve exponential equations. (pp. 501, 502)

Material Covered:

____ Example 1: Solving by Equating Exponents

____ Example 2: Taking a Logarithm of Each Side

____ Example 3: Taking a Logarithm of Each Side

____ Example 4: Using an Exponential Model

____ **Goal 2:** Solve logarithmic equations. (pp. 503, 504)

Material Covered:

____ Example 5: Solving a Logarithmic Equation

____ Example 6: Exponentiating Each Side

____ Student Help: Look Back

____ Example 7: Checking for Extraneous Solutions

____ Example 8: Using a Logarithmic Model

____ Other (specify) ____

____

Homework and Additional Learning Support

____ Textbook (specify) pp. 505–508

____

____ Internet: Extra Examples at www.mcdougallittell.com

____ *Reteaching with Practice* worksheet (specify exercises) ____________

____ *Personal Student Tutor* for Lesson 8.6

NAME ______________________ DATE ________

Interdisciplinary Application

For use with pages 501–508

The pH system

CHEMISTRY The pH system was developed by a Danish biochemist, Soren Sorensen in 1909. The symbol pH stands for "potential hydrogen" and is used to describe the concentration of hydrogen ions in a particular solution. A neutral solution has a pH of 7, acidic solutions have a pH below 7, and basic solutions have a pH above 7. Distilled water, for example, has a pH of 7, while human blood under normal conditions should have a pH between 7.35–7.45.

pH can be measured with an electronic pH meter or by using special dyes called acid-base indicators. The concentration of hydrogen ions determines the color of the indicator. Litmus paper is an example of an acid-base indicator that when dipped in a solution, turns red if the solution is an acid, or blue if the solution is a base.

In Exercises 1–4, use the following information.

The pH of a person's blood can be determined using the Henderson-Hasselbach equation, which is

$$\text{pH} = 6.1 + \log_{10}\left(\frac{x}{y}\right),$$

where x is the concentration of bicarbonate and y is the concentration of carbonic acid.

1. Expand the Henderson-Hasselbach equation.
2. Calculate the pH of a person's blood if the concentration of bicarbonate is 25.4 and the concentration of carbonic acid is 1.2. Round your result to three decimal places. Is the person's pH normal?
3. A person's blood has a pH of 7.371 and a carbonic acid concentration of 1.5. Approximate the concentration of bicarbonate in the person's blood. Round your result to two decimal places.
4. A person's blood has a pH of 7.408 and a bicarbonate concentration of 26.4. Approximate the concentration of carbonic acid in the person's blood. Round your result to two decimal places.

NAME ______________________ DATE __________

Challenge: Skills and Applications

For use with pages 501–508

1. a. Show that, for any positive numbers x and a, with $x \neq 1$, $\log_{x^2} a = \frac{1}{2} \log_x a$.

   b. Generalize the assertion in part (a) to powers other than x^2.

In Exercises 2–7, solve the equation.

2. $\dfrac{5^{x^2}}{25^x} = 125$

3. $\dfrac{(\sqrt{3})^{x+2}}{27^x} = 3^{-x^2}$

4. $\log_4 x - \log_{16}(x + 3) = \frac{1}{2}$

5. $\log(x + 12) + \log(x - 3) = 2$

6. $\log_x 2 + \log_x 3 = 5$

7. $\log_{x^2} 9 - \log_x 36 = 1$

8. The "Rule of 70" gives the approximate time T that it takes for an amount of money invested at an annual interest rate of $r\%$ to double in value:

   $$T = \frac{70}{r}.$$

   a. Write the compound interest formula for a principal P, an annual interest rate r, and a time T, with $2P$ as the final value of the investment.

   b. Solve the equation you wrote in part (a) for T by taking "ln" of both sides. Use the approximation: $\ln(1 + r) \approx r$, valid for small values of r.

   c. Explain why the Rule of 70 works.

9. The *half-life* of an unstable isotope is the amount of time it takes for half of any amount of the substance to decay.

   a. Express, as an exponential function, the amount A of an isotope remaining, in terms of the original amount present A_0, the time t over which the decay has taken place, and the half-life h. Use $\frac{1}{2}$ as the base of the exponential function.

   b. Neptunium-232 is an unstable isotope of Neptunium-228, which decays spontaneously to the more stable form. Suppose that after 60 minutes, 5.9% of the original amount of this isotope remains. Find the half-life of Neptunium-232.

LESSON **8.6**

NAME ______________________ DATE __________

Quiz 2

For use after Lessons 8.4–8.6

Evaluate the expression without using a calculator. *(Lesson 8.4)*

1. $\log_2 16$ **2.** $\log_5 0.04$ **3.** $\log_9 3$

Graph the function. State the domain and range. *(Lesson 8.5)*

4. $y = 2 + \log_3 x$ **5.** $y = \log_3(x - 2)$

Use a property of logarithms to evaluate the expression. *(Lesson 8.5)*

6. $\log_4(16 \cdot 4)$ **7.** $\log_3 \frac{1}{3}$

8. Expand the expression $\log_5 x^{\frac{1}{3}}y^6$. *(Lesson 8.5)*

9. Condense the expression $3 \log_3 15 + 2 \log_3 x - \log_3 25$. *(Lesson 8.5)*

10. Use the change-of-base formula to evaluate the expression $\log_5 23$. *(Lesson 8.5)*

Solve the equation. *(Lesson 8.6)*

11. $4e^x - 2 = 14$ **12.** $5 \log_2 x = 24$

13. $\ln(3x + 1) = \ln(2x - 8)$ **14.** $\log(x + 2) = -\log(x - 1) + 1$

15. ***Account Balance*** You deposit \$1000 in an account at 5% interest compounded continuously. How long will it take your money to double? *(Lesson 8.6)*

Answers

1. ______
2. ______
3. ______
4. Use grid at left.
5. Use grid at left.
6. ______
7. ______
8. ______
9. ______
10. ______
11. ______
12. ______
13. ______
14. ______
15. ______

LESSON 8.7

TEACHER'S NAME ______________________ CLASS ______ ROOM ______ DATE ______

Lesson Plan

1-day lesson (See *Pacing the Chapter,* TE pages 462C–462D) **For use with pages 509–516**

GOALS
1. **Model data with exponential functions.**
2. **Model data with power functions.**

State/Local Objectives ______________________

✓ **Check the items you wish to use for this lesson.**

STARTING OPTIONS

____ Homework Check: TE page 505; Answer Transparencies
____ Warm-Up or Daily Homework Quiz: TE pages 509 and 508, CRB page 94, or Transparencies

TEACHING OPTIONS

____ Lesson Opener (Visual Approach): CRB page 95 or Transparencies
____ Graphing Calculator Activity with Keystrokes: CRB page 96
____ Examples 1–6: SE pages 509–512
____ Extra Examples: TE pages 510–512 or Transparencies
____ Closure Question: TE page 512
____ Guided Practice Exercises: SE page 513

Lesson 8.7

APPLY/HOMEWORK

Homework Assignment

____ Basic 18–26 even, 30–38 even, 42–50 even, 55–59 odd, 61–71 odd
____ Average 18–38 even, 42–54 even, 55–59 odd, 61–79 odd
____ Advanced 18–38 even, 42–54 even, 55–60, 61–79 odd

Reteaching the Lesson

____ Practice Masters: CRB pages 97–99 (Level A, Level B, Level C)
____ Reteaching with Practice: CRB pages 100–101 or Practice Workbook with Examples
____ Personal Student Tutor

Extending the Lesson

____ Applications (Real-Life): CRB page 103
____ Challenge: SE page 516; CRB page 104 or Internet

ASSESSMENT OPTIONS

____ Checkpoint Exercises: TE pages 510–512 or Transparencies
____ Daily Homework Quiz (8.7): TE page 516, CRB page 107, or Transparencies
____ Standardized Test Practice: SE page 516; TE page 516; STP Workbook; Transparencies

Notes ______________________

LESSON 8.7

TEACHER'S NAME ______________________ CLASS ______ ROOM ______ DATE ______

Lesson Plan for Block Scheduling

Half-day lesson (See *Pacing the Chapter,* TE pages 462C–462D) **For use with pages 509–516**

1. **Model data with exponential functions.**
2. **Model data with power functions.**

State/Local Objectives ______________________________________

CHAPTER PACING GUIDE	
Day	**Lesson**
1	8.1 (all); 8.2(all)
2	8.3 (all)
3	8.4 (all)
4	8.5 (all)
5	8.6 (all)
6	**8.7 (all)**; 8.8(all)
7	Review/Assess Ch. 8

✓ Check the items you wish to use for this lesson.

STARTING OPTIONS

____ Homework Check: TE page 505; Answer Transparencies
____ Warm-Up or Daily Homework Quiz: TE pages 509 and 508, CRB page 94, or Transparencies

TEACHING OPTIONS

____ Lesson Opener (Visual Approach): CRB page 95 or Transparencies
____ Graphing Calculator Activity with Keystrokes: CRB page 96
____ Examples 1–6: SE pages 509–512
____ Extra Examples: TE pages 510–512 or Transparencies
____ Closure Question: TE page 512
____ Guided Practice Exercises: SE page 513

APPLY/HOMEWORK

Homework Assignment (See also the assignment for Lesson 8.8.)

____ Block Schedule: 18–38 even, 42–54 even, 55–59 odd, 61–79 odd

Reteaching the Lesson

____ Practice Masters: CRB pages 97–99 (Level A, Level B, Level C)
____ Reteaching with Practice: CRB pages 100–101 or Practice Workbook with Examples
____ Personal Student Tutor

Extending the Lesson

____ Applications (Real Life): CRB page 103
____ Challenge: SE page 516; CRB page 104 or Internet

ASSESSMENT OPTIONS

____ Checkpoint Exercises: TE pages 510–512 or Transparencies
____ Daily Homework Quiz (8.7): TE page 516, CRB page 107, or Transparencies
____ Standardized Test Practice: SE page 516; TE page 516; STP Workbook; Transparencies

Notes ______________________________________

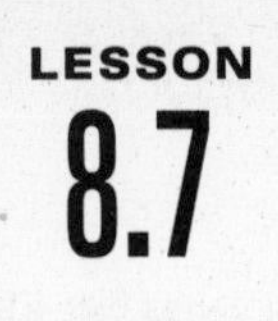

Available as a transparency

NAME ______________________ DATE __________

WARM-UP EXERCISES

For use before Lesson 8.7, pages 509–516

1. The graph of $y = 2 \cdot 5^x$ passes through $(x_1, 250)$ and $(2, y_2)$. Find the values of x_1 and y_2.

2. What is the general form of an exponential equation?

3. Write $2.7 = 4^b$ in logarithmic form.

4. Use the properties of exponents to simplify $4^{0.5x+3}$.

5. Evaluate $\dfrac{\log 2.5}{\log 0.7}$.

Lesson 8.7

DAILY HOMEWORK QUIZ

For use after Lesson 8.6, pages 501–508

Tell whether the *x*-value is a solution of the equation.

1. $\log_6 4x = 3;\ x = 0.25 \cdot 6^3$

2. $2e^x - 3 = 9;\ x = \ln 12$

Solve the equation.

3. $3^{2x} = 27^{x+2}$

4. $5e^{3x} + 2 = 17$

Solve the equation. Check for extraneous solutions.

5. $\log_4(5x - 11) = \log_4(3 - 2x)$

6. $-3 \ln \dfrac{x}{2} = 4$

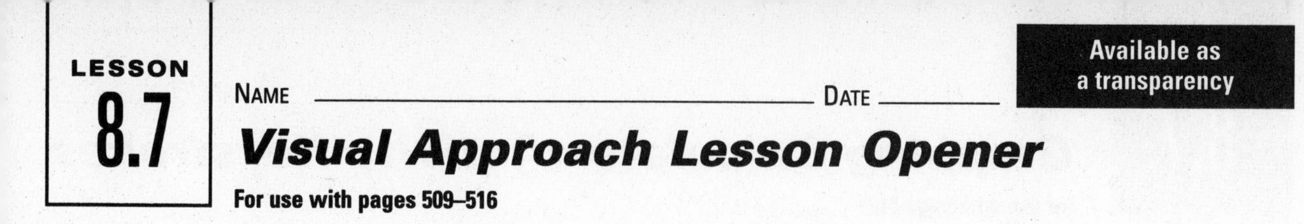

LESSON 8.7

NAME ______________________________ DATE ____________

Available as a transparency

Visual Approach Lesson Opener

For use with pages 509–516

If $y = f(x)$ is an exponential function in the form $f(x) = ab^x$, then the graph of $y = \ln f(x)$ has a special property.

Graph $y = f(x)$ and $y = \ln f(x)$ on the same set of axes.

1. $f(x) = 3^x$

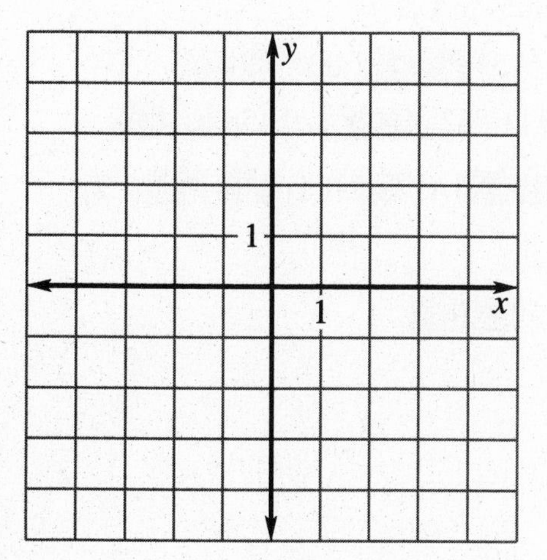

2. $f(x) = \left(\frac{1}{2}\right)\left(\frac{2}{3}\right)^x$

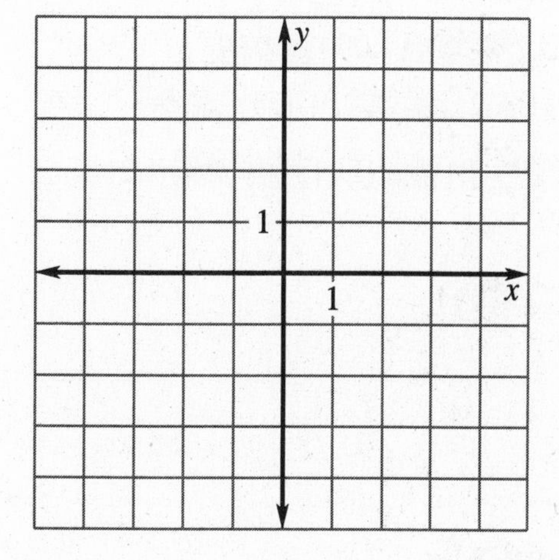

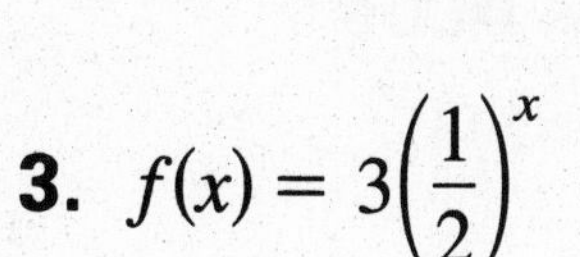

3. $f(x) = 3\left(\frac{1}{2}\right)^x$

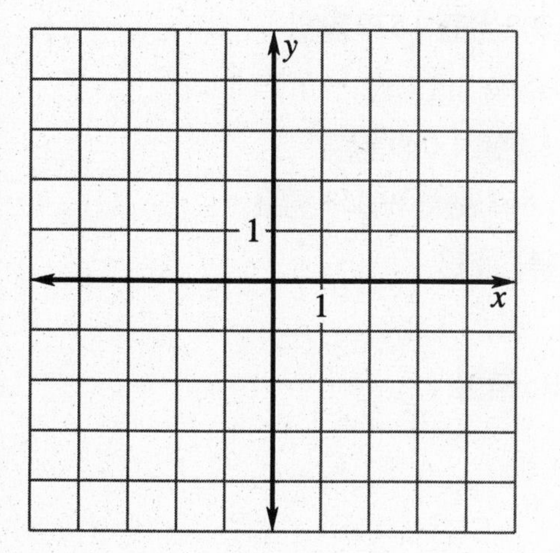

4. $f(x) = \frac{2}{3} \cdot 2^x$

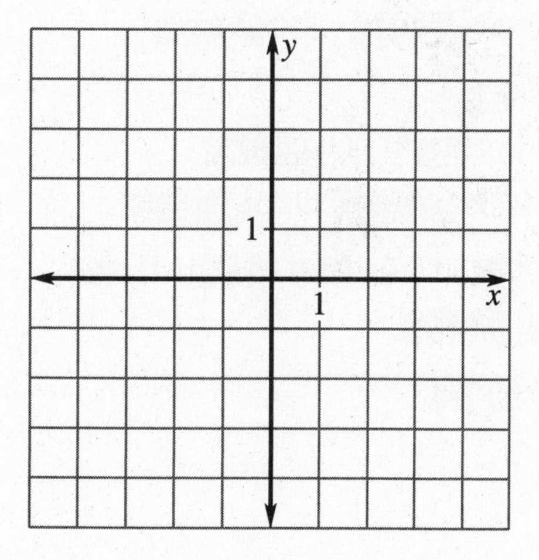

5. If $y = f(x)$ is an exponential function, what type of function is $y = \ln f(x)$?

NAME ______________________ DATE __________

Graphing Calculator Activity Keystrokes

For use with page 512

Keystrokes for Example 6

TI-82

STAT 1

Enter the x-values in L1.

.387 ENTER .723 ENTER 1.000 ENTER

1.524 ENTER 5.203 ENTER 9.539 ENTER

Enter the y-values in L2.

.241 ENTER .615 ENTER 1.000 ENTER

1.881 ENTER 11.862 ENTER 29.458 ENTER

STAT ▶ ALPHA [B] 2nd [L1] , 2nd [L2]

ENTER

30.043 ^ 1.5 ENTER

TI-83

STAT 1

Enter the x-values in L1.

.387 ENTER .723 ENTER 1.000 ENTER

1.524 ENTER 5.203 ENTER 9.539 ENTER

Enter the y-values in L2.

.241 ENTER .615 ENTER 1.000 ENTER

1.881 ENTER 11.862 ENTER 29.458 ENTER

STAT ▶ ALPHA [A] 2nd [L1] , 2nd [L2]

ENTER

30.043 ^ 1.5 ENTER

SHARP EL-9600c

STAT [A] ENTER

Enter the x-values in L1.

.387 ENTER .723 ENTER 1.000 ENTER

1.524 ENTER 5.203 ENTER 9.539 ENTER

Enter the y-values in L2.

.241 ENTER .615 ENTER 1.000 ENTER

1.881 ENTER 11.862 ENTER 29.458 ENTER

2ndF [QUIT] STAT [D] 0 9 (2ndF [L1] ,

2ndF [L2]) ENTER

30.043 a^b 1.5 ENTER

CASIO CFX-9850GA PLUS

From the main menu, choose STAT.

Enter the x-values in List 1.

.387 EXE .723 EXE 1.000 EXE

1.524 EXE 5.203 EXE 9.539 EXE

Enter the y-values in List 2.

.241 EXE .615 EXE 1.000 EXE

1.881 EXE 11.862 EXE 29.458 EXE

F2 F3 F6 F3

MENU 1

30.043 ^ 1.5 EXE

NAME ______________________________ DATE __________

Practice A

For use with pages 509–516

Write an exponential function of the form $y = ab^x$ whose graph passes through the given points.

1. (0, 1), (3, 27)
2. (1, 6), (2, 12)
3. (1, 10), (2, 50)
4. (1, 2), (2, 8)
5. (4, 16), (6, 64)
6. (2, 18), (3, 54)

Use the table of values to determine whether or not an exponential model is a good fit for the data (t, y).

7.

t	1	2	3	4	5	6	7	8
$\ln y$	0.23	0.64	1.07	1.47	1.88	2.31	2.72	3.12

8.

t	1	2	3	4	5	6	7	8
$\ln y$	1.32	1.52	1.92	2.72	2.88	3.52	4.32	5.6

9.

t	1	2	3	4	5	6	7	8
$\ln y$	0.05	0.17	0.27	0.40	0.52	0.63	0.75	0.85

10.

t	1	2	3	4	5	6	7	8
$\ln y$	12.31	13.56	14.82	16.04	17.29	18.49	19.76	21.01

Solve for y.

11. $\ln y = 0.324t + 1.601$
12. $\ln y = 1.203t + 0.418$
13. $\ln y = 12.135t + 5.144$
14. $\ln y = 3.207t + 1.132$
15. $\ln y = 1.032t + 8.149$
16. $\ln y = 2.301t + 1.624$

Write a power function of the form $y = ax^b$ whose graph passes through the given points.

17. (1, 2), (3, 54)
18. (1, 3), (2, 12)
19. (1, 1), (4, 8)

Use the table of values to determine whether or not a power function model is a good fit for the data (x, y).

20.

$\ln x$	0	0.693	1.099	1.386	1.609
$\ln y$	1.264	2.594	3.924	5.254	6.584

21.

$\ln x$	0	0.693	1.099	1.386	1.609
$\ln y$	0.833	2.219	3.030	3.605	4.052

Solve for y.

22. $\ln y = 2.4 \ln x$
23. $\ln y = 1.3 \ln x$
24. $\ln y = 0.8 \ln x$

NAME ______________________________ DATE __________

Practice B

For use with pages 509–516

Write an exponential function of the form $y = ab^x$ whose graph passes through the given points.

1. $\left(1, \frac{2}{3}\right), \left(2, \frac{4}{3}\right)$ **2.** $\left(2, \frac{16}{25}\right), \left(3, \frac{64}{25}\right)$ **3.** $\left(2, \frac{3}{4}\right), \left(3, \frac{3}{8}\right)$

4. $(1, 4), \left(2, \frac{8}{3}\right)$ **5.** $\left(1, \frac{5}{2}\right), \left(2, \frac{25}{2}\right)$ **6.** $\left(2, \frac{5}{36}\right), \left(3, \frac{5}{108}\right)$

Use the table of values to draw a scatter plot of ln y versus x. Then find an exponential model for the data.

7.

x	1	2	3	4	5	6	7	8
y	8	16	32	64	128	256	512	1024

8.

x	1	2	3	4	5	6	7	8
y	3.6	8.64	20.736	49.766	119.439	286.654	687.971	1651.13

9.

x	1	2	3	4	5	6	7	8
y	3	4.5	6.75	10.125	15.188	22.781	34.172	51.258

Write a power function of the form $y = ax^b$ whose graph passes through the given points.

10. (2, 16), (3, 36) **11.** (2, 4), (4, 32) **12.** (2, 64), (3, 486)

13. (4, 9.6), (9, 32.4) **14.** (4, 384), (16, 49,152) **15.** (2, 9.879), (3, 16.070)

Use the table of values to draw a scatter plot of ln y versus ln x. Then find a power model for the data.

16.

x	1	2	3	4	5	6	7	8
y	1.5	6	13.5	24	37.5	54	73.5	96

17.

x	1	2	3	4	5	6	7	8
y	2.4	7.275	13.919	22.055	31.518	42.194	53.997	66.858

18. ***Consumer Magazines*** The table shows the circulation of the top 10 consumer magazines in 1997 where x represents the magazine's ranking. Use a graphing calculator to find a power model for the data. Use the model to estimate the circulation of the 15th ranked magazine.

Rank	*Circulation (millions)*	*Rank*	*Circulation (millions)*
1	20.454	6	7.615
2	20.432	7	5.054
3	15.086	8	4.643
4	13.171	9	4.514
5	9.013	10	4.256

NAME ______________________________ DATE __________

Practice C

For use with pages 509–516

Write an exponential function of the form $y = ab^x$ whose graph passes through the given points.

1. (2, 5.888), (3, 9.4208) **2.** (1, 0.9), (2, 0.18) **3.** (2, 41.552), (3, 116.3456)

Find an exponential model for the data.

4.

x	1	2	3	4	5	6	7	8
y	14	49	171.5	600.25	2100.9	7353.1	25,736	90,075

5.

x	1	2	3	4	5	6	7	8
y	3	6	12	24	48	96	192	384

6. ***Critical Thinking*** To determine whether an exponential model fits the data, you need to determine whether the data of the form $(x, \ln y)$ is linear. To see that this test works, start with $y = ab^x$, take the natural logarithm of both sides, and use the properties of logarithms to verify that there is a linear relationship between x and $\ln y$.

Write a power function of the form $y = ax^b$ whose graph passes through the given points.

7. (4, 3), (9, 4.5) **8.** (4, 19.2), (9, 64.8) **9.** (16, 16.6), (81, 24.9)

Find a power model for the data.

10.

x	1	2	3	4	5	6	7
y	3	6.8922	11.212	15.834	20.696	25.757	30.991

11.

x	1	2	3	4	5	6	7
y	2.5	14.142	38.971	80	139.75	220.45	324.1

12. ***Critical Thinking*** To determine whether a power model fits the data, you need to determine whether the data of the form $(\ln x, \ln y)$ is linear. To see that this test works, start with $y = ax^b$, take the natural logarithm of both sides, and use the properties of logarithms to verify that there is a linear relationship between $\ln x$ and $\ln y$.

Volunteer Work **In Exercises 13–15, use the following information.**

The table below shows the percent of the adult population P that participates in volunteer work as a function of household income where $t = 1$ represents a household income under \$10,000, $t = 2$ represents a household income between \$10,000 and \$19,000, and so on.

t	1	2	3	4	5	6
P	34.7	34.3	41.2	46.0	52.7	64.1

13. Use your graphing calculator to find an exponential model for the data.

14. Use your graphing calculator to find a power model for the data.

15. Which model is the better fitting model? Explain your answer.

LESSON
8.7

NAME ______________________ DATE __________

Reteaching with Practice

For use with pages 509–516

GOAL **Model data with exponential functions and power functions**

EXAMPLE 1 *Writing an Exponential Function*

Write an exponential function $y = ab^x$ whose graph passes through $(2, -36)$ and $(0, -4)$.

SOLUTION

Begin by substituting the coordinates of the two given points to obtain two equations in a and b.

$-36 = ab^2$ Substitute -36 for y and 2 for x.

$-4 = ab^0$ Substitute -4 for y and 0 for x.

Notice that the second equation becomes $-4 = a$ because $b^0 = 1$. Substitute $a = -4$ in the first equation and solve for b:

$-36 = (\mathbf{-4})b^2$ Substitute -4 for a.

$9 = b^2$ Divide each side by -4.

$3 = b$ Take the positive square root.

So, $y = -4 \cdot 3^x$.

Exercises for Example 1

Write an exponential function $y = ab^x$ whose graph passes through the given points.

1. $(0, 7), (1, 14)$ **2.** $(1, -12), (-1, -3)$ **3.** $(1, 9), (-1, 1)$

Lesson 8.7

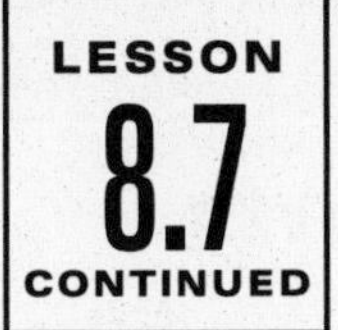

Reteaching with Practice

For use with pages 509–516

EXAMPLE 2 Writing a Power Function

Write a power function $y = ax^b$ whose graph passes through (2, 4) and (4, 32).

SOLUTION

Begin by substituting the coordinates of the two points to obtain two equations in a and b.

$4 = a \cdot 2^b$ Substitute 4 for y and 2 for x.

$32 = a \cdot 4^b$ Substitute 32 for y and 4 for x.

To solve the system, solve for a in the first equation to get $a = \frac{4}{2^b}$, then substitute into the second equation.

$$32 = \left(\frac{4}{2^b}\right)4^b$$

$$32 = 4 \cdot 2^b$$

$$8 = 2^b$$

By inspection, $b = 3$, so $a = \frac{4}{2^b} = \frac{4}{2^3} = \frac{4}{8} = 0.5$ and $y = 0.5x^3$.

Exercises for Example 2

Write a power function of the form $y = ax^b$ whose graph passes through the given points.

4. (2, 1), (6, 9) **5.** (4, 48), (2, 6) **6.** (9, 6), (4, 4)

NAME ______________________________ DATE ________

Quick Catch-Up for Absent Students

For use with pages 509–516

The items checked below were covered in class on (date missed) __________

Lesson 8.7: Modeling with Exponential and Power Functions

____ **Goal 1:** Model data with exponential functions. (pp. 509, 510)

Material Covered:

____ Example 1: Writing an Exponential Function

____ Student Help: Look Back

____ Example 2: Finding an Exponential Model

____ Example 3: Using Exponential Regression

____ **Goal 2:** Model data with power functions. (pp. 511, 512)

Material Covered:

____ Example 4: Writing a Power Function

____ Example 5: Finding a Power Model

____ Example 6: Using Power Regression

____ Other (specify) ______________________________

____

Homework and Additional Learning Support

____ Textbook (specify) pp. 513–516

____

____ *Reteaching with Practice* worksheet (specify exercises) ______________

____ *Personal Student Tutor* for Lesson 8.7

NAME ______________________________ DATE __________

Real-Life Application: When Will I Ever Use This?

For use with pages 509–516

Professional Basketball

The table below shows the rank x and points per game p of the top scorers in the National Basketball Association for the 1998-1999 season. Each player's team is in parentheses.

Player (Team)	*Rank, x*	*Points per game, p*
Allen Iverson (Philadelphia)	1	26.8
Shaquille O'Neal (Los Angeles Lakers)	2	26.3
Karl Malone (Utah)	3	23.8
Shareef Abdur-Rahim (Vancouver)	4	23.0
Keith Van Horn (New Jersey)	5	21.8
Tim Duncan (San Antonio)	6	21.7
Gary Payton (Seattle)	7	21.7
Stephon Marbury (New Jersey)	8	21.3
Antonio McDyess (Denver)	9	21.2
Grant Hill (Detroit)	10	21.1
Kevin Garnett (Minnesota)	11	20.8
Shawn Kemp (Cleveland)	12	20.5

1. Draw a scatter plot of $\ln p$ versus x.

2. Draw a scatter plot of $\ln p$ versus $\ln x$.

3. Using the scatter plots from Exercises 1 and 2, determine whether an exponential model or a power model is a better fit for the original data. Explain.

4. Use your answer from Exercise 3 to find a model for the data. Check your model by using the regression feature of a graphing calculator.

5. Use the model that you found in Exercise 4 with your graphing calculator to complete the table shown below.

Player (Team)	*Rank, x*	*Points per game, p*
Michael Finley (Dallas)	13	
Alonzo Mourning (Miami)	14	
Kobe Bryant (Los Angeles Lakers)	15	
Mitch Richmond (Washington)	16	

Lesson 8.7

NAME ______________________ DATE __________

Challenge: Skills and Applications

For use with pages 509–516

Solve each equation.

1. $(\ln x)^2 - \ln x - 6 = 0$
2. $e^{-2x} + 3e^{-x} = 10$
3. Given an exponential equation, $y = ab^x$, with $b > 0$ and $b \neq 1$, you can write it in the form $y = ac^{kx}$, for any given positive number c, $c \neq 1$.
   a. By substituting the expression for y in one of these two equations into the other, express k in terms of b and c.
   b. Let $f(x) = 100\left(\frac{2}{3}\right)^x$. Express $f(x)$ using base e. Express $f(x)$ using base $\frac{1}{2}$.
4. The slope m of the tangent line to the graph of the exponential function $y = ab^x$ at the point x_0 is given by

   $m = (\ln b)ab^{x_0}$.

   a. Show that the slope of the tangent line at the point x_0 equals $f(x_0)$ if and only if $b = e$.
   b. Suppose the tangent line to the graph of $y = ab^x$ has slope 4 at $x = 0$, and has slope 36 at $x = 2$. Find an equation of the exponential function.
5. a. Solve the equation

   $\ln y = x^2 \ln b + \ln a$

   for y.

   b. The *standard normal curve* is the graph of a function $f(x)$ of the form you found in part (a) which passes through the points

   $$\left(0, \frac{1}{\sqrt{2\pi}}\right) \text{ and } \left(1, \frac{1}{\sqrt{2\pi e}}\right).$$

   Find an equation of the standard normal curve.

   c. What can you say about the function $f(x)$ whose graph is the standard normal curve, as $x \to \infty$? As $x \to -\infty$? Where does the graph have a turning point?

Lesson 8.7

LESSON 8.8

TEACHER'S NAME ______________________ CLASS ________ ROOM ________ DATE ________

Lesson Plan

1-day lesson (See *Pacing the Chapter,* TE pages 462C–462D) **For use with pages 517–522**

GOALS

1. Evaluate and graph logistic growth functions.
2. Use logistic growth functions to model real-life quantities.

State/Local Objectives ____

____

✓ Check the items you wish to use for this lesson.

STARTING OPTIONS

____ Homework Check: TE page 513; Answer Transparencies
____ Warm-Up or Daily Homework Quiz: TE pages 517 and 516, CRB page 107, or Transparencies

TEACHING OPTIONS

____ Motivating the Lesson: TE page 518
____ Lesson Opener (Graphing Calculator): CRB page 108 or Transparencies
____ Graphing Calculator Activity with Keystrokes: CRB page 109
____ Examples 1–5: SE pages 517–519
____ Extra Examples: TE pages 518–519 or Transparencies; Internet
____ Closure Question: TE page 519
____ Guided Practice Exercises: SE page 520

APPLY/HOMEWORK

Homework Assignment

____ Basic 16–26, 27–41 odd, 52, 55–63 odd; Quiz 3: 1–7
____ Average 16–26, 27–43 odd, 50–52, 55–63 odd; Quiz 3: 1–7
____ Advanced 16–26, 27–43 odd, 45–47, 50–53, 55–63 odd; Quiz 3: 1–7

Reteaching the Lesson

____ Practice Masters: CRB pages 110–112 (Level A, Level B, Level C)
____ Reteaching with Practice: CRB pages 113–114 or Practice Workbook with Examples
____ Personal Student Tutor

Extending the Lesson

____ Applications (Interdisciplinary): CRB page 116
____ Challenge: SE page 522; CRB page 117 or Internet

ASSESSMENT OPTIONS

____ Checkpoint Exercises: TE pages 518–519 or Transparencies
____ Daily Homework Quiz (8.8): TE page 522 or Transparencies
____ Standardized Test Practice: SE page 522; TE page 522; STP Workbook; Transparencies
____ Quiz (8.7–8.8): SE page 522

Notes ____

____

____

LESSON 8.8

TEACHER'S NAME ______________________ CLASS ______ ROOM ______ DATE ______

Lesson Plan for Block Scheduling

Half-day lesson (See *Pacing the Chapter,* TE pages 462C–462D) For use with pages 517–522

GOALS

1. Evaluate and graph logistic growth functions.
2. Use logistic growth functions to model real-life quantities.

State/Local Objectives ______________________

CHAPTER PACING GUIDE	
Day	**Lesson**
1	8.1 (all); 8.2(all)
2	8.3 (all)
3	8.4 (all)
4	8.5 (all)
5	8.6 (all)
6	8.7 (all); **8.8(all)**
7	Review/Assess Ch. 8

✓ Check the items you wish to use for this lesson.

STARTING OPTIONS

____ Homework Check: TE page 513; Answer Transparencies
____ Warm-Up or Daily Homework Quiz: TE pages 517 and 516, CRB page 107, or Transparencies

TEACHING OPTIONS

____ Motivating the Lesson: TE page 518
____ Lesson Opener (Graphing Calculator): CRB page 108 or Transparencies
____ Graphing Calculator Activity with Keystrokes: CRB page 109
____ Examples 1–5: SE pages 517–519
____ Extra Examples: TE pages 518–519 or Transparencies; Internet
____ Closure Question: TE page 519
____ Guided Practice Exercises: SE page 520

APPLY/HOMEWORK

Homework Assignment (See also the assignment for Lesson 8.7.)

____ Block Schedule: 16–26, 27–43 odd, 50–52, 55–63 odd; Quiz 3: 1–7

Reteaching the Lesson

____ Practice Masters: CRB pages 110–112 (Level A, Level B, Level C)
____ Reteaching with Practice: CRB pages 113–114 or Practice Workbook with Examples
____ Personal Student Tutor

Extending the Lesson

____ Applications (Interdisciplinary): CRB page 116
____ Challenge: SE page 522; CRB page 117 or Internet

ASSESSMENT OPTIONS

____ Checkpoint Exercises: TE pages 518–519 or Transparencies
____ Daily Homework Quiz (8.8): TE page 522 or Transparencies
____ Standardized Test Practice: SE page 522; TE page 522; STP Workbook; Transparencies
____ Quiz (8.7–8.8): SE page 522

Notes ______________________

LESSON 8.8

NAME ______________________ DATE ________

Available as a transparency

WARM-UP EXERCISES

For use before Lesson 8.8, pages 517–522

Simplify.

1. $\dfrac{\ln 3}{0.36}$

2. $\dfrac{\ln 10}{10}$

3. Evaluate $f(4)$ for $f(x) = 1 + 2e^{-3x}$.

4. What is the horizontal asymptote of the graph of $y = 1 + 2e^{-2x}$?

5. Solve $7 = 35e^{-4x}$.

DAILY HOMEWORK QUIZ

For use after Lesson 8.7, pages 509–516

1. Write an exponential function $y = ab^x$ whose graph passes through $(-2, 81)$ and $(2, 16)$.

2. Find the ordered pairs $(x, \ln y)$ for the data. Then find an exponential model for the data. (1, 7.20), (2, 12.96), (3, 23.33), (4, 41.99), (5, 75.58).

3. Write a power function $y = ax^b$ whose graph passes through $(1, 2)$ and $(4, 16)$.

4. Draw a scatter plot of the ordered pairs $(\ln x, \ln y)$ for the data. Then find a power model for the data. (2, 12.25), (3, 31.64), (4, 62.03), (5, 104.57), (6, 160.21), (7, 229.80)

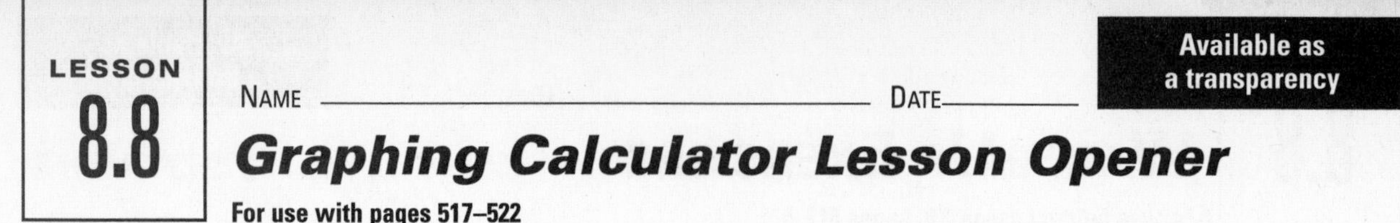

LESSON 8.8

NAME ______________________ DATE ____________

Available as a transparency

Graphing Calculator Lesson Opener

For use with pages 517–522

Suppose that a population of 25 rabbits is introduced to a small island. As time goes on, the rabbit population increases. The table below gives population data.

Year (t)	0	1	2	3	4	5	6	7	8	9
Population (P)	25	31	33	41	44	55	62	71	77	82

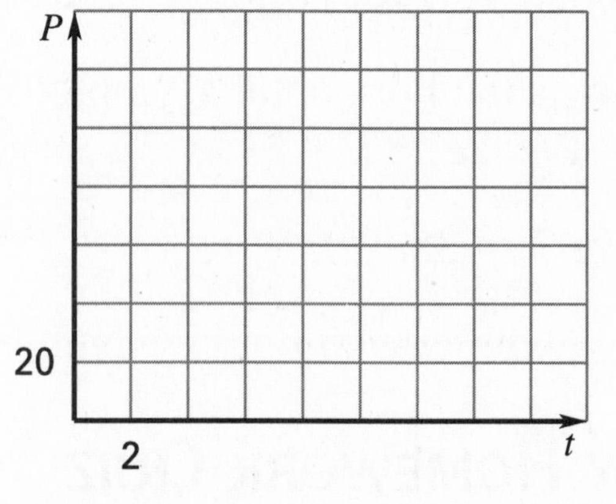

1. Use exponential regression to find an exponential model for the data.
2. Sketch a scatter plot of the data along with the exponential model.
3. Use the model to estimate the population for $t = 20$, $t = 30$, and $t = 50$.
4. In Lesson 8.8, you will learn about *logistic growth functions*. Use a graphing calculator to find a logistic model for the data.
5. Add a graph of the logistic model to your scatter plot.
6. Use the logistic model to estimate the population for $t = 20$, $t = 30$, and $t = 50$.
7. Which model do you think will be a better model for the long-term rabbit population growth on the island? Explain your answer.

NAME ______________________ DATE __________

Graphing Calculator Activity Keystrokes

For use with page 519

Keystrokes for Example 5

TI-82

The TI-82 can do regression in general, but it cannot do logistic regression.

TI-83

STAT 1

Enter the x-values in L1.

0 ENTER 1 ENTER 2 ENTER 3 ENTER

4 ENTER 5 ENTER 6 ENTER 7 ENTER

8 ENTER 9 ENTER 10 ENTER

Enter the y-values in L2.

18 ENTER 33 ENTER 56 ENTER 90 ENTER

130 ENTER 170 ENTER 203 ENTER 225

ENTER 239 ENTER 247 ENTER 251 ENTER

STAT ▶ ALPHA [B] 2nd [L1] , 2nd [L2]

ENTER

SHARP EL-9600C

STAT [A] ENTER

Enter the x-values in L1.

0 ENTER 1 ENTER 2 ENTER 3 ENTER

4 ENTER 5 ENTER 6 ENTER 7 ENTER

8 ENTER 9 ENTER 10 ENTER

Enter the y-values in L2.

18 ENTER 33 ENTER 56 ENTER 90 ENTER

130 ENTER 170 ENTER 203 ENTER 225

ENTER 239 ENTER 247 ENTER 251 ENTER

2ndF [QUIT] STAT [D] 1 3 (2ndF [L1] ,

2ndF [L2]) ENTER

CASIO CFX-9850GA PLUS

From the main menu, choose STAT.

Enter the x-values in List 1.

0 EXE 1 EXE 2 EXE 3 EXE 4 EXE 5 EXE

6 EXE 7 EXE 8 EXE 9 EXE 10 EXE

Enter the y-values in List 2.

18 EXE 33 EXE 56 EXE 90 EXE

130 EXE 170 EXE 203 EXE 225 EXE

239 EXE 247 EXE 251 EXE

F2 F3 F6 F5

LESSON **8.8**

NAME ______________________ DATE ________

Practice A

For use with pages 517–522

Evaluate the function $f(x) = \frac{2}{1 + e^{-x}}$ for the given value of x.

1. $f(1)$
2. $f(-1)$
3. $f(6)$
4. $f(0)$
5. $f\left(\frac{1}{2}\right)$
6. $f(3.4)$
7. $f(-0.2)$
8. $f\left(\frac{5}{4}\right)$

Match the function with its graph.

9. $f(x) = \frac{3}{1 + e^{-x}}$
10. $f(x) = \frac{3}{1 + e^{-2x}}$
11. $f(x) = \frac{1}{1 + 2e^{-x}}$

A.
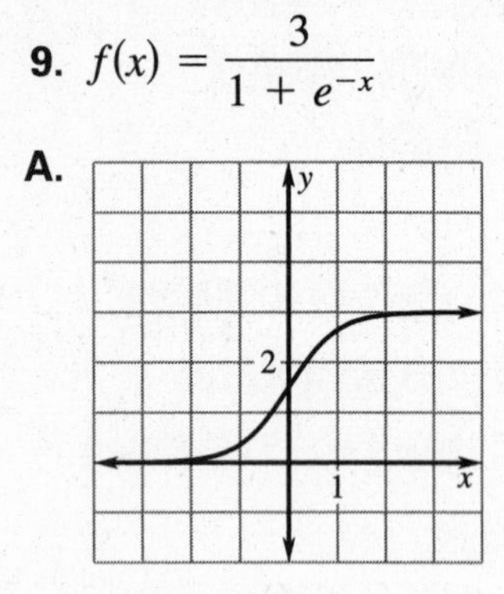

B.
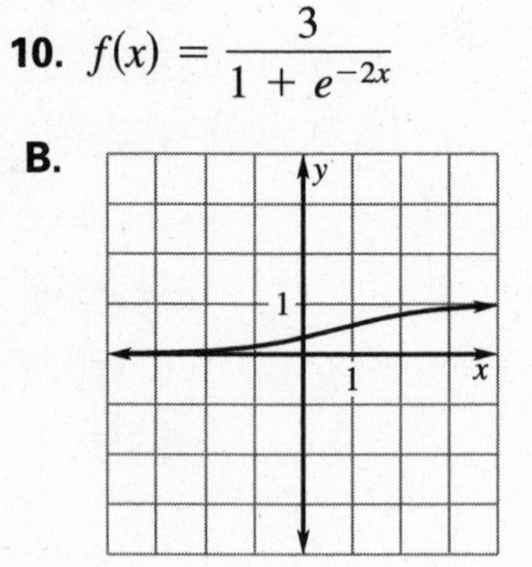

C.
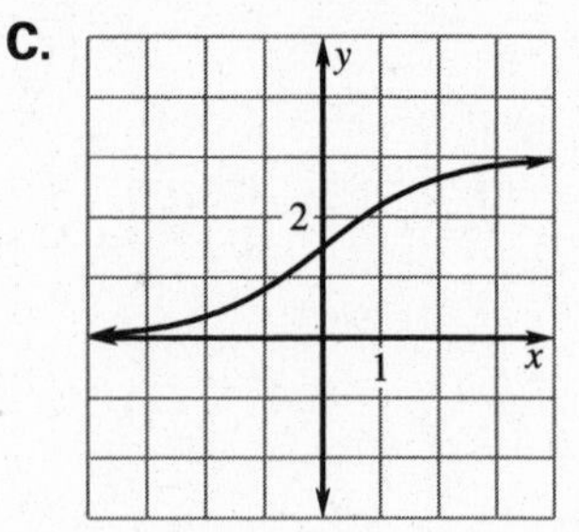

Identify the horizontal asymptotes of the function.

12. $f(x) = \frac{1}{1 + 4e^{-2x}}$
13. $f(x) = \frac{5}{1 + e^{-2x}}$
14. $f(x) = \frac{6}{1 + 2e^{-x}}$

Identify the y-intercept of the function.

15. $y = \frac{1}{1 + 2e^{-x}}$
16. $y = \frac{4}{1 + e^{-x}}$
17. $y = \frac{5}{1 + e^{-3x}}$

Identify the point of maximum growth of the function.

18. $f(x) = \frac{4}{1 + e^{-2x}}$
19. $f(x) = \frac{1}{1 + 3e^{-x}}$
20. $f(x) = \frac{2}{1 + 2e^{-3x}}$

Advertising **In Exercises 21 and 22, use the following information.**

A company decides to stop advertising one of its products. The sales of the product S can be modeled by

$$S = \frac{100{,}000}{1 + 0.5e^{-0.3t}}$$

where t is the number of years since advertising stopped.

21. What are the sales 5 years after advertising stopped?
22. What can the company expect in terms of sales in the future?

NAME ______________________________ DATE __________

Practice B

For use with pages 517–522

Tell whether the function is an example of *exponential growth, exponential decay, logarithmic,* or *logistics growth*.

1. $f(x) = \left(\frac{1}{2}\right)^x$
2. $f(x) = \ln 3x$
3. $f(x) = \frac{1}{1 + 3e^{-x}}$

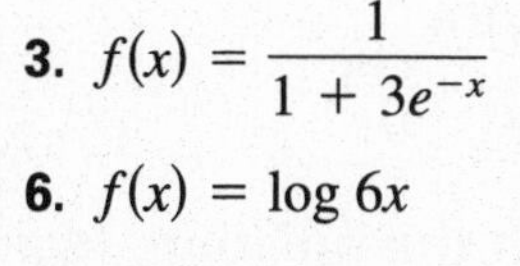

4. $f(x) = e^{-2x}$
5. $f(x) = 2.5^x$
6. $f(x) = \log 6x$

Match the function with its graph.

7. $f(x) = \frac{4}{1 + 2e^{-x}}$
8. $f(x) = \frac{2}{1 + 2e^{-x}}$
9. $f(x) = \frac{4}{1 + e^{-2x}}$

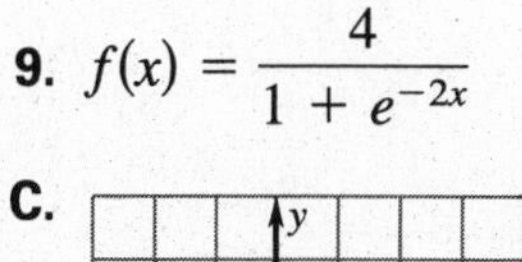

A.

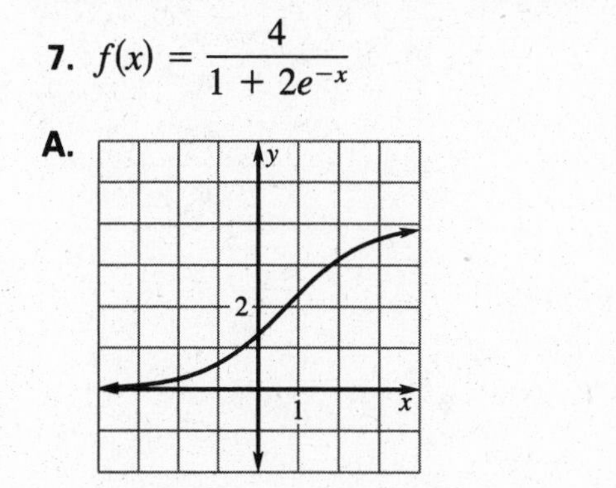

B.

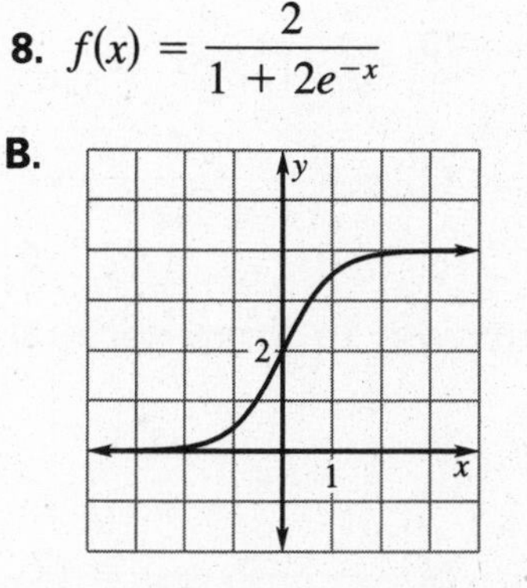

C.

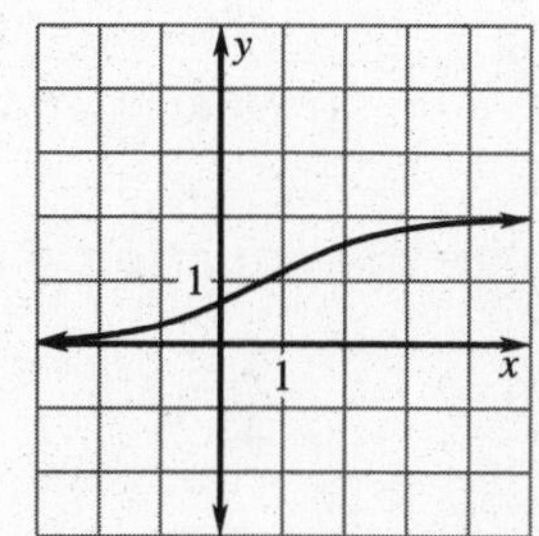

Identify the horizontal asymptotes of the function.

10. $f(x) = \frac{20}{1 + 0.4e^{-x}}$
11. $f(x) = -5 + \frac{1}{1 + e^{-x}}$
12. $f(x) = 10 + \frac{2}{1 + e^{-x}}$

Sketch the graph of the function.

13. $f(x) = \frac{3}{1 + e^{-x}}$
14. $f(x) = \frac{1}{1 + 5e^{-x}}$
15. $f(x) = 1 + \frac{5}{1 + e^{-x}}$

Solve the equation.

16. $\frac{4}{1 + 2e^{-x}} = 2$
17. $\frac{8}{1 + e^{-x}} = 5$
18. $\frac{12}{1 + 5e^{-2x}} = 6$

Wildlife Management **In Exercises 19–22, use the following information.**

A wildlife organization releases 100 deer into a wilderness area. The deer population P can be modeled by

$$P = \frac{500}{1 + 4e^{-0.36t}}$$

where t is the time in years.

19. Sketch the graph of the model.
20. Identify the horizontal asymptotes of the graph.
21. What is the maximum number of deer the wilderness area can support?
22. What is the deer population after 10 years?

LESSON 8.8

NAME ______________________ DATE __________

Practice C

For use with pages 517–522

Evaluate the function $f(x) = \dfrac{8}{1 + 2e^{-3x}}$ for the given value of x.

1. $f(0)$
2. $f(1)$
3. $f(-1)$
4. $f(5)$
5. $f(-5)$
6. $f\left(\frac{2}{5}\right)$
7. $f\left(-\frac{3}{4}\right)$
8. $f\left(\frac{7}{3}\right)$

Graph the function. Identify the asymptotes, y-intercept, and point of maximum growth.

9. $y = \dfrac{2}{1 + e^{-x}}$
10. $y = \dfrac{4}{1 + 2e^{-3x}}$
11. $y = \dfrac{7}{1 + 3e^{-2x}}$
12. $y = \dfrac{3}{1 + 4e^{-5x}}$
13. $y = \dfrac{25}{1 + 12e^{-2x}}$
14. $y = \dfrac{10}{1 + 5e^{-1/2x}}$

Solve the equation. Round the answer to three decimal places.

15. $\dfrac{10}{1 + 2e^{-x}} = 5$
16. $\dfrac{12}{1 + 3e^{-4x}} = 3$
17. $\dfrac{13}{1 + 6e^{-2x}} = 5$
18. $\dfrac{28}{1 + 6e^{-3x}} = 14$
19. $\dfrac{32}{1 + 4.5e^{-2.5x}} = 18$
20. $\dfrac{40}{1 + 8e^{-0.8x}} = 8.5$

Conservation **In Exercises 21–23, use the following information.**

A conservation organization believes that the growth of a population P of an endangered species at its preserve can be modeled by the curve

$$P = \frac{2000}{1 + 10e^{kt}}$$

where t is time in years.

21. After 1 year, the preserve's population of endangered species is 215. Find k.
22. When will the population reach 500?
23. What is the maximum population the preserve can maintain?
24. ***Analyzing Models*** The graph of the logistic growth function $y = \dfrac{c}{1 + ae^{-rx}}$ has a y-intercept of $\dfrac{c}{1 + a}$. Verify this formula by setting x equal to 0 and solving for y.

Analyzing Models **In Exercises 25–29, use the function $y = \dfrac{c}{1 + ae^{-rx}}$.**

25. As $x \to \infty$, what is the behavior of $-rx$?
26. As $x \to \infty$, what is the behavior of e^{-rx}?
27. As $x \to \infty$, what is the behavior of ae^{-rx}?
28. As $x \to \infty$, what is the behavior of $1 + ae^{-rx}$?
29. As $x \to \infty$, what is the behavior of $\dfrac{c}{1 + ae^{-rx}}$

Lesson 8.8

LESSON 8.8

NAME ______________________________ DATE __________

Reteaching with Practice

For use with pages 517–522

GOAL Evaluate and graph logistic growth functions

Vocabulary

Logistic growth functions are written as $y = \dfrac{c}{1 + ae^{-rx}}$, where c, a, and r are positive constants.

The graph of $y = \dfrac{c}{1 + ae^{-rx}}$ has the following characteristics:

- The horizontal lines $y = 0$ and $y = c$ are asymptotes.
- The y-intercept is $\dfrac{c}{1 + a}$.
- The domain is all real numbers, and the range is $0 < y < c$.
- The graph is increasing from left to right. To the left of its point of maximum growth, $\left(\dfrac{\ln a}{r}, \dfrac{c}{2}\right)$, the rate of increase is increasing. To the right of its point of maximum growth, the rate of increase is decreasing.

EXAMPLE 1 *Evaluating a Logistic Growth Function*

Evaluate $f(x) = \dfrac{300}{1 + e^{-2x}}$ for (a) $f(-2)$, (b) $f(0)$, and (c) $f(3)$.

Solution

a. $f(-2) = \dfrac{300}{1 + e^{-2(-2)}} = \dfrac{300}{1 + e^4} \approx 5.4$

b. $f(0) = \dfrac{300}{1 + e^{-2(0)}} = \dfrac{300}{1 + e^0} = \dfrac{300}{1 + 1} = 150$

c. $f(3) = \dfrac{300}{1 + e^{-2(3)}} = \dfrac{300}{1 + e^{-6}} \approx 299.3$

Exercises for Example 1

Evaluate the function $f(x) = \dfrac{5}{1 + e^{-0.3x}}$ for the given value of x.

1. $f(0)$ **2.** $f(1)$ **3.** $f(-1)$

4. $f(4)$ **5.** $f(-3)$ **6.** $f(0.6)$

EXAMPLE 2 *Graphing a Logistic Growth Function*

Graph $y = \dfrac{2}{1 + 3e^{-x}}$.

Lesson 8.8

NAME ______________________ DATE __________

Reteaching with Practice

For use with pages 517–522

SOLUTION

Begin by sketching the horizontal asymptote, $y = 2$. Then find the y-intercept at $y = \frac{2}{1 + 3} = 0.5$. The point of maximum growth is $\left(\frac{\ln 3}{1}, \frac{2}{2}\right) \approx (1.1, 1)$. Plot these points. Finally, from left to right, draw a curve that starts just above the x-axis, curves up to the point of maximum growth, and then levels off as it approaches the upper horizontal asymptote, $y = 2$.

Exercises for Example 2

Graph the function. Identify the asymptotes, *y*-intercept, and point of maximum growth.

7. $y = \frac{4}{1 + 3e^{-x}}$　　**8.** $y = \frac{3}{1 + e^{-0.02x}}$　　**9.** $y = \frac{2}{1 + 2e^{-3x}}$

EXAMPLE 3 Solving a Logistic Growth Equation

Solve $\frac{12}{1 + 3e^{-2x}} = 10$.

SOLUTION

$\frac{12}{1 + 3e^{-2x}} = 10$	Write original equation.
$12 = 10(1 + 3e^{-2x})$	Multiply each side by $1 + 3e^{-2x}$.
$12 = 10 + 30e^{-2x}$	Use distributive property.
$2 = 30e^{-2x}$	Subtract 10 from each side.
$0.067 = e^{-2x}$	Divide each side by 30.
$\ln 0.067 = \ln e^{-2x}$	Take natural log of each side.
$\ln 0.067 = -2x$	$\ln e^x = x$
$-\frac{1}{2} \ln 0.067 = x$	Multiply each side by $-\frac{1}{2}$.
$1.35 \approx x$	Use a calculator.

The solution is about 1.35.

Exercises for Example 3

Solve the equation.

10. $\frac{25}{1 + 2e^{-x}} = 20$　　**11.** $\frac{4}{1 + e^{-4x}} = 1$　　**12.** $\frac{100}{1 + 5e^{-3x}} = 50$

Lesson 8.8

NAME ______________________________ DATE __________

Quick Catch-Up for Absent Students

For use with pages 517–522

The items checked below were covered in class on (date missed) __________

Lesson 8.8: Logistic Growth Functions

____ **Goal 1:** Evaluate and graph logistic growth functions. (pp. 517, 518)

Material Covered:

____ Example 1: Evaluating a Logistic Growth Function

____ Activity: Graphs of Logistic Growth Functions

____ Example 2: Graphing a Logistic Growth Function

____ Example 3: Solving a Logistic Growth Equation

Vocabulary:

logistic growth functions, p. 517

____ **Goal 2:** Use logistic growth functions to model real-life quantities. (p. 519)

Material Covered:

____ Example 4: Using a Logistic Growth Model

____ Student Help: Keystroke Help

____ Example 5: Writing a Logistic Growth Model

____ Other (specify) ______________________________

____

Homework and Additional Learning Support

____ Textbook (specify) pp. 520–522

____

____ Internet: Extra Examples at www.mcdougallittell.com

____ *Reteaching with Practice* worksheet (specify exercises) __________

____ *Personal Student Tutor* for Lesson 8.8

NAME ______________________ DATE ________

Interdisciplinary Application

For use with pages 517–522

Endangered Species

BIOLOGY The term "endangered" is used to define plants and animals that are currently in danger of becoming extinct. Species today become extinct primarily as a result of human activities such as overhunting, habitat destruction, and wildlife collecting. Organizations publish lists of endangered species so that lawmakers, conservationists, and the public will take notice. But despite these efforts, many species of animals and plants become extinct every year. Some examples of animals that are currently on the endangered species list are the giant panda, the red wolf, the ivory-billed woodpecker, and the snow leopard.

The Endangered Species Act of 1973 was passed to help protect endangered wildlife from becoming extinct. This act not only reshaped the way people view plants and animals, but it made clear that it is our responsibility to maintain the survival of these species. Since the act was passed, endangered species such as the bald eagle and peregrine falcon have increased in certain areas.

Research has shown just how much species depend on one another and how much we depend on certain species. It is important for people to think of the long-term effects of eliminating a species instead of our short-term needs. The effects of any species becoming extinct can be immeasurable, and once a species is lost, it can never be brought back.

In Exercises 1–4, use the following information.

A conservation organization releases 100 animals of an endangered species into a game preserve. The organization believes that the preserve can hold up to 1000 animals, and that the endangered species population P will be modeled by

$$P(t) = \frac{1000}{1 + 9e^{-0.1656t}},$$

where t is the time measured in months.

1. Graph the function. What are the horizontal asymptotes of the graph?
2. What will the endangered species population be after 8 months? after 14 months?
3. After how many months will the population be 650?
4. Suppose the table below shows the population of the endangered species through 14 months. Use a graphing calculator to find a logistic growth model that gives P as function of t.

t	0	1	2	3	4	5	6	7
P	100	114	132	156	180	225	281	349

t	8	9	10	11	12	13	14
P	390	453	502	542	568	597	620

NAME ______________________________ DATE ______________

Challenge: Skills and Applications

For use with pages 517–522

1. Find the logistic function whose graph passes through the point (1, 2) and has point of maximum growth (2, 4).

2. **a.** For the logistic function you found in Exercise 1, show that the point (3, 6) is also on the graph.

   b. Explain what the result of part (a) indicates about the symmetry of the graph.

3. The symmetry property mentioned in Exercise 2(b) can be expressed more generally for the graph of the logistic function $f(x) = \dfrac{c}{1 + e^{-rx}}$ by saying that

   $f(x) - \dfrac{c}{2} = \dfrac{c}{2} - f(-x),$

   for all x. Show that this relationship holds by computing both sides separately.

4. The growth rate R of a logistic function

   $$y = \frac{c}{1 + ae^{-rx}}$$

   at a point (x, y) can be given by the formula

   $$R = \frac{r}{c}y(c - y).$$

   a. Show that the growth rate approaches 0 as y approaches c and also as y approaches 0.

   b. Show that the maximum value of the growth rate occurs at $y = \dfrac{c}{2}$.

5. If you suspend a flexible string (or chain) by two points at the same height above the ground, the string falls naturally into a shape called a *catenary*, whose graph (in the simplest case) is given by a function called *cosh* (pronounced "kosh"). A similar function, called *sinh* (pronounced "cinch") is naturally associated with the cosh function in many ways. The two functions are given by the following equations.

   $\cosh(x) = \frac{1}{2}(e^x + e^{-x})$ $\sinh(x) = \frac{1}{2}(e^x - e^{-x})$

   a. Show that, for any x, $[\cosh(x)]^2 - [\sinh(x)]^2 = 1$.

   b. Show that, for any x,

   (i) $\sinh(2x) = 2\cosh(x)\sinh(x)$, and

   (ii) $\cosh(2x) = [\cosh(x)]^2 + [\sinh(x)]^2$.

NAME ______________________________ DATE __________

Chapter Review Games and Activities

For use after Chapter 8

Match the graph with the type of function by placing the appropriate lower case letter in the blank in front of the numbered name of the function. Match the equations with a capital letter in the blank behind the name of the function. You must use all graphs. One of the functions will have two answers.

_____ **1.** Exponential growth _____

_____ **2.** Exponential decay _____

_____ **3.** Logistic growth _____

_____ **4.** Logarithmic _____

a. $y = \left(\frac{1}{2}\right)3^x$

b. $y = -5\left(\frac{2}{3}\right)^x$

c. $y = 4\left(\frac{2}{3}\right)^x$

d. $y = \ln(x - 2)$

e. $y = \frac{5}{(1 + 5e^{-x})}$

A.

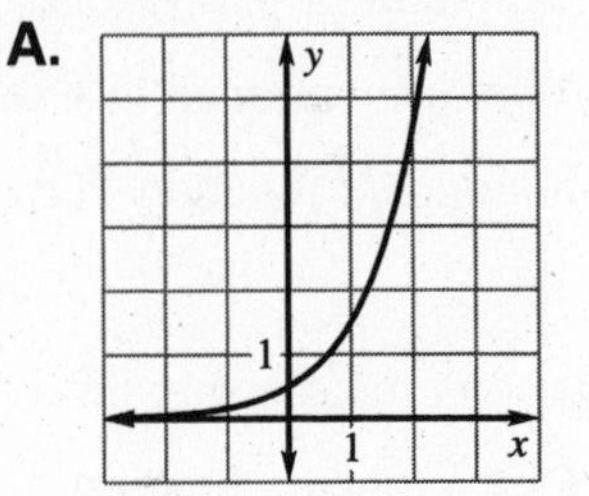

B.

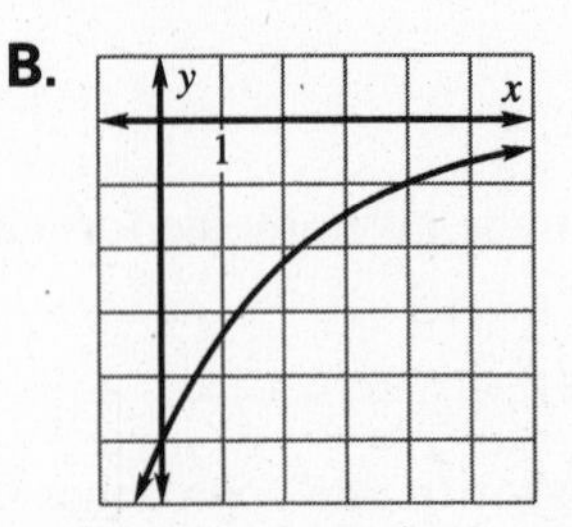

C.

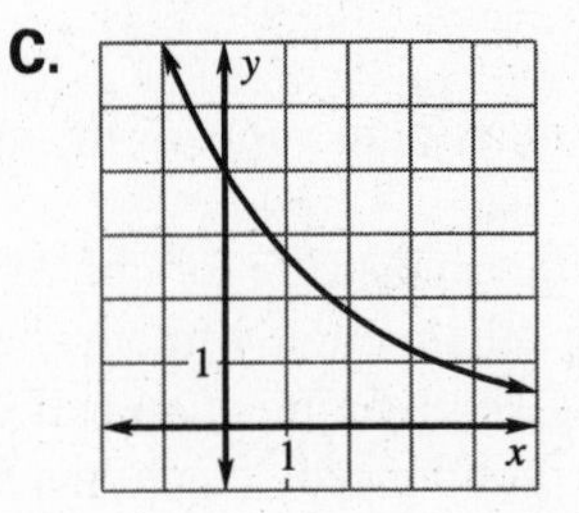

D.

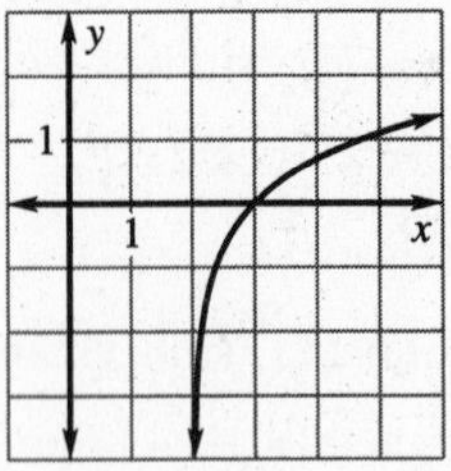

E.

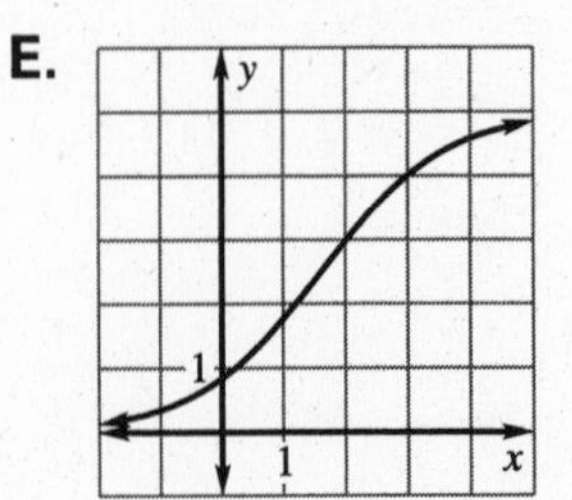

NAME ______________________ DATE ____________

Chapter Test A

For use after Chapter 8

Graph the function. State the domain and range.

1. $y = 2^x$

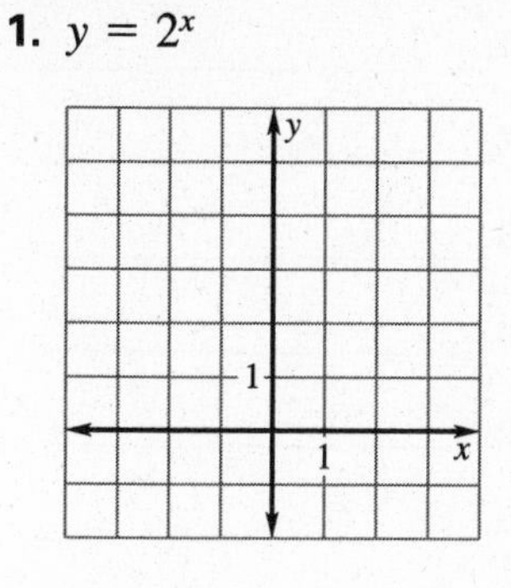

2. $y = 2^{(x-1)} + 1$

3. $y = \frac{1}{3}e^x$

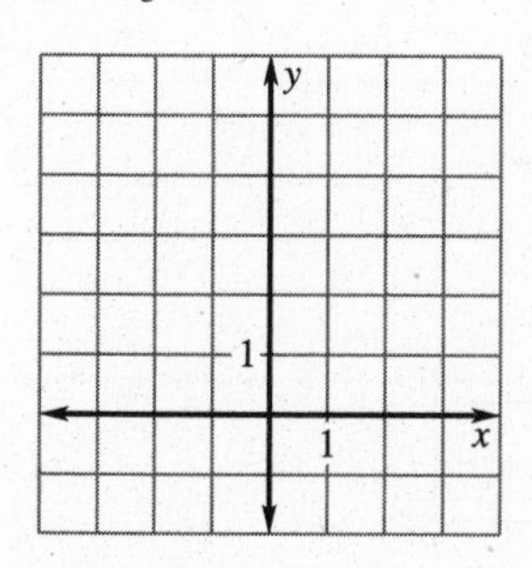

4. $y = \log x$

5. $y = \ln(x - 1)$

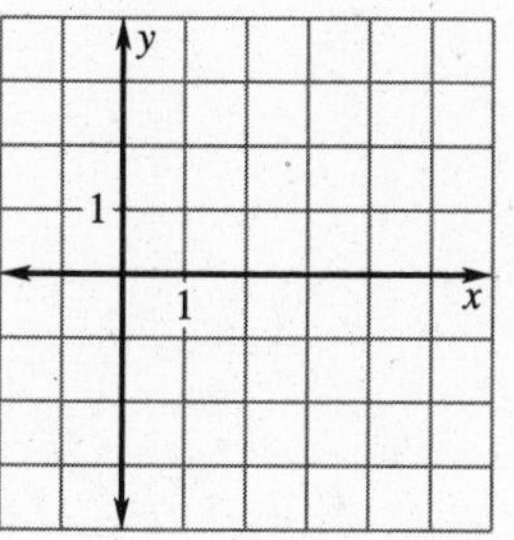

6. $y = 2e^{-x}$

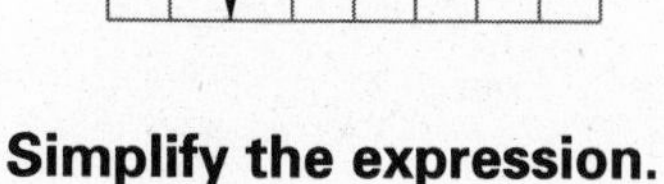

Simplify the expression.

7. $(e^3)(e^2)$

8. $(3e)(e^{-2})$

9. $\log 10{,}000$

10. $\log_3 27$

11. $\frac{e^4}{e^3} \cdot \frac{-3}{e}$

Evaluate the expression without using a calculator.

12. $\log_2 0.5$

13. $\log_{1/2} 4$

14. $\log_3 1$

15. $\ln e^1$

Solve the equation. Check for extraneous solutions.

16. $10^{3x+5} = 10^{x-3}$

17. $\log_3(2x - 1) = 2$

18. $\log_5(4x + 1) = \log_5(2x + 7)$

19. $\log_2(y + 4) + \log_2 y = 5$

Answers

1. Use grid at left.
2. Use grid at left.
3. Use grid at left.
4. Use grid at left.
5. Use grid at left.
6. Use grid at left.
7. ____________
8. ____________
9. ____________
10. ____________
11. ____________
12. ____________
13. ____________
14. ____________
15. ____________
16. ____________
17. ____________
18. ____________
19. ____________

Review and Assess

NAME ______________________ DATE __________

Chapter Test A

For use after Chapter 8

20. Tell whether the function $f(x) = 4\left(\frac{3}{2}\right)^x$ represents *exponential growth* or *exponential decay.*

21. Find the inverse of the function $y = \log_5 x$.

Use log 5 ≈ 0.699 and log 12 ≈ 1.079 to approximate the value of the expression.

22. $\log 25$

23. $\log \frac{1}{12}$

24. Condense the expression $3 \log x + \log 7$.

25. Expand the expression $\ln 3xy$.

26. Use the change-of-base formula to evaluate the expression $\log_5 10$.

27. Find an exponential function of the form $y = ab^x$ whose graph passes through the points (2, 1) and (3, 2).

28. Find a power function of the form $y = ax^b$ whose graph passes through the points (4, 4) and (16, 8).

29. ***Car Depreciation*** The value of a new car purchased for $20,000 decreases by 10% per year. Write an exponential decay model for the value of the car. Use the model to estimate the value after one year.

30. ***Earning Interest*** You deposit $1000 in an account that pays 6% annual interest compounded continuously. Find the balance at the end of 2 years.

20. ______
21. ______
22. ______
23. ______
24. ______
25. ______
26. ______
27. ______
28. ______
29. ______
30. ______

NAME ______________________ DATE __________

Chapter Test B

For use after Chapter 8

Graph the function. State the domain and range.

1. $y = 3^x$

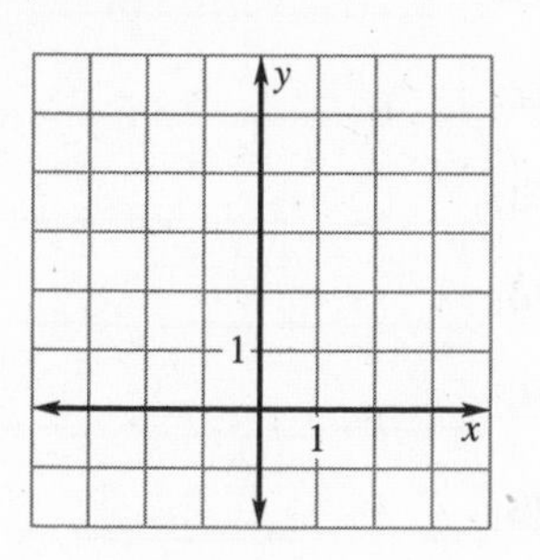

2. $y = 2 \log x$

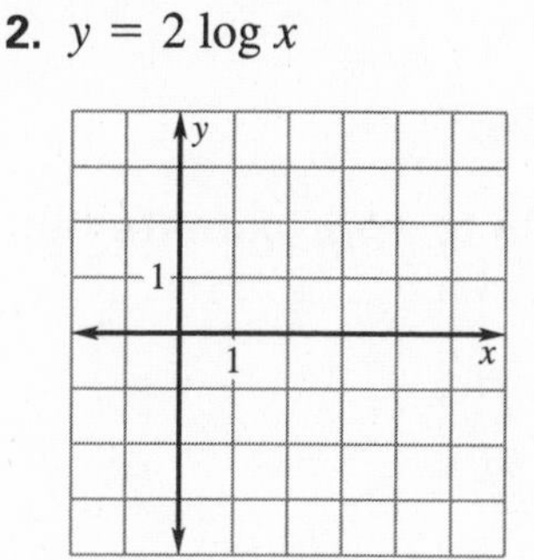

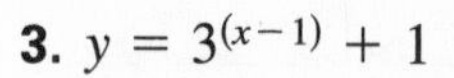

3. $y = 3^{(x-1)} + 1$

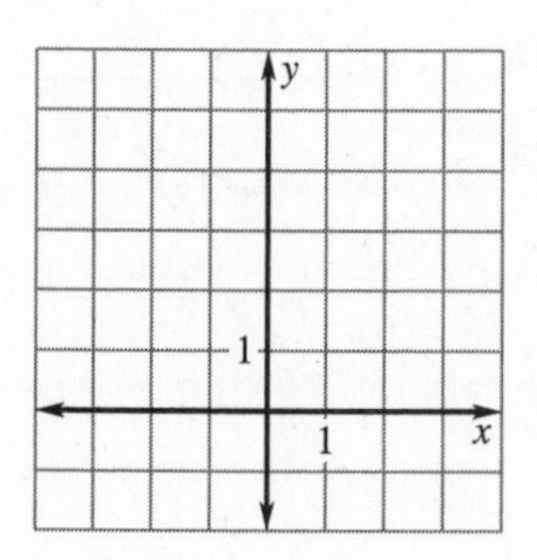

4. $y = \frac{2}{3}e^x$

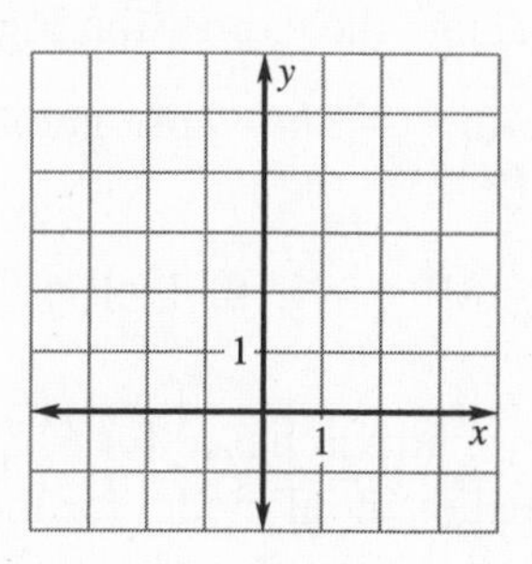

5. $y = 3^x - 3$

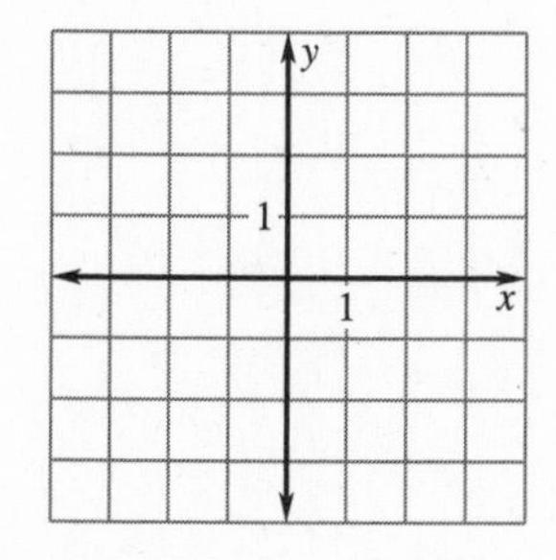

6. $y = \dfrac{4}{1 + 2e^{-x}}$

Simplify the expression.

7. $(-e^2)(e^{-1})$
8. $(2e)(4e)(-e^{-3})$
9. $\log \frac{1}{100}$
10. $\log_5 3125$
11. $\dfrac{e^3 \cdot e^2 \cdot e}{e^{-1}}$

Evaluate the expression without using a calculator.

12. $\log_2 0.0625$
13. $\log_{1/2} 16$
14. $\log_{12} 1$
15. $\ln e^3$

Solve the equation. Check for extraneous solutions.

16. $\log_4 x = 2$
17. $10^{4x-1} = 1000$
18. $2e^x - 1 = 9$
19. $2 \log_5 x - \log_5 2 = \log_5(2x + 6)$

Answers

1. Use grid at left.
2. Use grid at left.
3. Use grid at left.
4. Use grid at left.
5. Use grid at left.
6. Use grid at left.
7. ______
8. ______
9. ______
10. ______
11. ______
12. ______
13. ______
14. ______
15. ______
16. ______
17. ______
18. ______
19. ______

NAME ______________________ DATE __________

Chapter Test B

For use after Chapter 8

20. Tell whether the function $y = \frac{1}{4}e^{2x}$ represents *exponential growth* or *exponential decay.*

21. Find the inverse of the function $y = \log_7 x$.

Use $\log_2 10 \approx 3.322$ and $\log 8 \approx 0.903$ to approximate the value of the expression.

22. $\log_2 100$

23. $\log \frac{1}{8}$

24. Condense the expression $\log 7 - \log b$.

25. Expand the expression $\ln \frac{5x}{2}$.

26. Use the change-of-base formula to evaluate the expression $\log_2 9$.

27. Find an exponential function of the form $y = ab^x$ whose graph passes through the points (1, 50) and (2, 25).

28. Find a power function of the form $y = ax^b$ whose graph passes through the points (2, 4) and (6, 8).

29. ***Car Depreciation*** The value of a new car purchased for \$18,000 decreases by 12% per year. Write an exponential model for the value of the car. Use the model to estimate the value after two years.

30. ***Earning interest*** You deposit 1000 in an account that pays 5% annual interest compounded continuously. Find the balance at the end of 3 years.

20. ______________
21. ______________
22. ______________
23. ______________
24. ______________
25. ______________
26. ______________
27. ______________
28. ______________
29. ______________
30. ______________

Review and Assess

NAME ______________________________ DATE __________

Chapter Test C

For use after Chapter 8

Graph the function. State the domain and range.

1. $y = \left(\frac{3}{2}\right)^x$

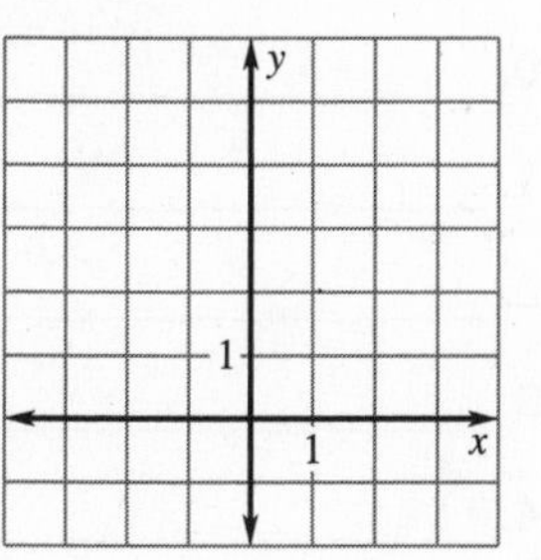

2. $y = \log_4 x$

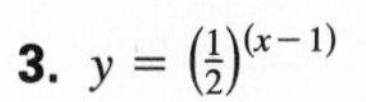

3. $y = \left(\frac{1}{2}\right)^{(x-1)}$

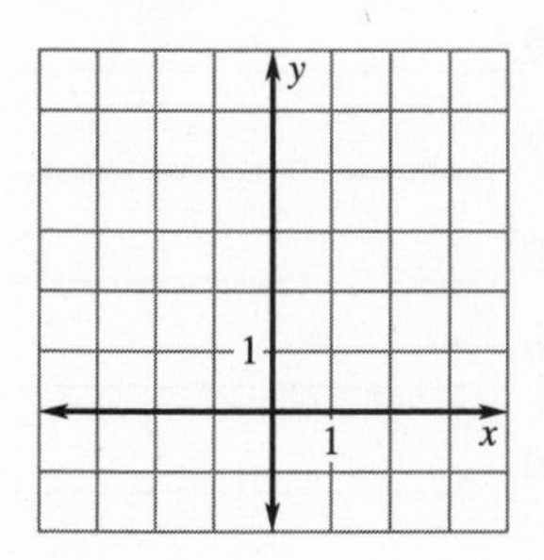

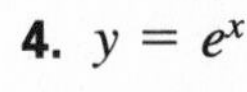

4. $y = e^x$

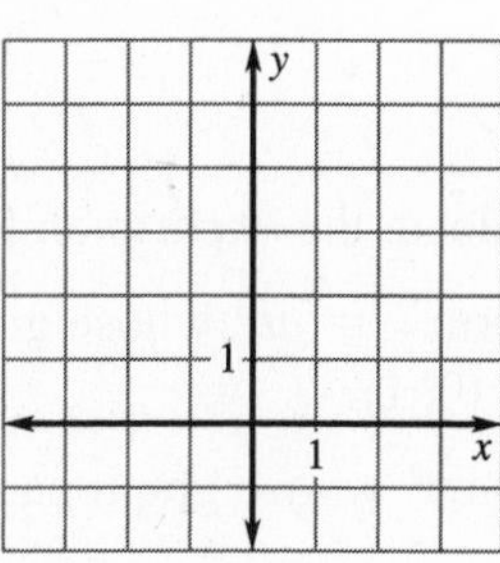

5. $y = -\ln x + 2$

6. $y = \dfrac{50}{1 + 125e^{-x}}$

Simplify the expression.

7. $(e)(e^3)$

8. $(3e)^2$

9. $\log \frac{1}{1000}$

10. $\log_2 32$

11. $\dfrac{4e^4}{e^5} \cdot \dfrac{e}{-2}$

Evaluate the expression without using a calculator.

12. $\log_2 0.25$

13. $\log_{1/2} 8$

14. $\log_2 1$

15. $\ln e^2$

Solve the equation. Check for extraneous solutions.

16. $\log_5 x = 4$

17. $10^{x^2+1} = 100{,}000$

Answers

1. Use grid at left. ____________

____

2. Use grid at left. ____________

____

3. Use grid at left. ____________

____

4. Use grid at left. ____________

____

5. Use grid at left. ____________

____

6. Use grid at left. ____________

____

7. ____________
8. ____________
9. ____________
10. ____________
11. ____________
12. ____________
13. ____________
14. ____________
15. ____________
16. ____________
17. ____________

Review and Assess

NAME ______________________ DATE __________

Chapter Test C

For use after Chapter 8

18. $2 \log_3 y = \log_3 4 + \log_3 (y + 8)$

19. $\ln(3x + 1) - \ln(x + 5) = 0$

20. Tell whether the function $f(x) = 3\left(\frac{1}{2}\right)^2$ represents *exponential growth* or *exponential decay.*

21. Find the inverse of the function $y = \log_8 x$.

Use $\log_8 100 \approx 2.214$ and $\log \frac{1}{15} \approx -1.176$ to approximate the value of the expression.

22. $\log_8 10{,}000$

23. $\log 15$

24. Condense the expression $\log_4 3 + 3 \log_4 2$.

25. Expand the expression $\ln \dfrac{2y}{x}$.

26. Use the change-of-base formula to evaluate the expression $\log_7 125$.

27. Find the exponential function of the form $y = ab^x$ whose graph passes through the points $\left(-3, \frac{1}{27}\right)$ and $(0, 1)$.

28. Find a power function of the form $y = ax^b$ whose graph passes through the points $(2, 5)$ and $(8, 12)$.

29. ***Car Depreciation*** The value of a new car purchased for \$28,000 decreases 8% per year. Write an exponential decay model for the value of the car. Use the model to estimate the value after 5 years.

30. ***Earning Interest*** You deposit \$800 in an account that pays $5\frac{1}{2}\%$ annual interest compounded continuously. Find the balance at the end of 5 years.

18. ______________

19. ______________

20. ______________

21. ______________

22. ______________

23. ______________

24. ______________

25. ______________

26. ______________

27. ______________

28. ______________

29. ______________

30. ______________

Review and Assess

NAME ______________________________ DATE __________

SAT/ACT Chapter Test

For use after Chapter 8

1. What is the log of 100 to the base 10?

Ⓐ 10 Ⓑ 2
Ⓒ 3 Ⓓ 1

2. The natural base e is

Ⓐ rational Ⓑ imaginary
Ⓒ irrational Ⓓ undefined

3. What is the simplified form of $\frac{8e^4}{-4e^3}$?

Ⓐ $-2e^7$ Ⓑ $-2e$
Ⓒ $\frac{1}{2e}$ Ⓓ $\frac{-e}{2}$

4. What type of function is $f(x) = 2e^{2x}$?

Ⓐ Exponential deacy function
Ⓑ Linear function
Ⓒ Quadratic function
Ⓓ Exponential growth

5. Which of the following is equivalent to $\log_5 5^3$?

Ⓐ -3 Ⓑ 125
Ⓒ 3 Ⓓ -125

6. What is the solution of the equation $9^{x+1} = 27^{x-1}$?

Ⓐ No solution Ⓑ -2
Ⓒ 5 Ⓓ 2

7. What is the asymptote of the graph of $f(x) = 2^x$?

Ⓐ x-axis Ⓑ y-axis
Ⓒ $y = 1$ Ⓓ $y = -1$

8. Which of the following is equivalent to $\log_b \frac{x}{y}$?

Ⓐ $\log_b x \div \log_b y$ Ⓑ $\log_b x - \log_b y$
Ⓒ $\log_b (x - y)^{1/2}$ Ⓓ $\log_b x + \log_b y$

Quantitative Comparision Exercises 9 and 10, choose the statement that is true about the given quantities.

Ⓐ The quantity in column A is greater.
Ⓑ The quantity in column B is greater.
Ⓒ The two quantities are equal.
Ⓓ The relationship cannot be determined from the given information.

9.

Column A	*Column B*
$\log_{10} 1000$	$\log_3 27$

Ⓐ Ⓑ Ⓒ Ⓓ

10.

Column A	*Column B*
The y value when $x = 0$ of the graph of $y = -2 \cdot 4^x$	The y value when $x = 0$ of the graph of $y = 2 \cdot 5^x$

Ⓐ Ⓑ Ⓒ Ⓓ

Review and Assess

NAME ______________________ DATE __________

Alternative Assessment and Math Journal

For use after Chapter 8

JOURNAL 1. **a.** Explain the difference in the graphs of $f(x) = a \cdot 3^x$ and $g(x) = -a\left(\frac{1}{3}\right)^x$, where $a > 0$.

b. Compare the graphs of $h(x) = -a \cdot 3^x$, $j(x) = a\left(\frac{1}{3}\right)^x$, and $y = \log_3 x$. (Assume $a > 0$.)

c. Which of the equations from parts a and b are inverses? Explain how this can be determined.

MULTI-STEP PROBLEM 2. Two neighboring towns have had population changes over a ten year period that follow exponential growth or exponential decay patterns.

- The population of Town A was 50,000 in 1980. It has increased in population by approximately 5.1% per year.
- The population of Town B was 100,000 in 1980. It has experienced a decrease in population of 8.1% per year.

a. Write an exponential model to describe the population of Town A. Estimate the population in the year 1988.

b. Write an exponential model to describe the population of Town B.

c. Estimate the population of town B in the year 2005. Is this a good approximation? Explain why or why not.

d. Which model represents exponential growth? Which is a model of exponential decay? Explain why exponential growth or exponential decay models can be used for this data.

e. Approximately how many years would it take the population of Town A to double? Determine the solution algebraically.

f. In how many years would the population of Town B decrease by 25%? Determine the solution by graphing.

3. ***Critical Thinking*** Determine the year when the towns in Exercise 2 would have approximately the same population. Use two different methods. Explain why using an algebraic method would be more difficult here than in Exercise 2, part d.

Alternative Assessment Rubric

For use after Chapter 8

JOURNAL SOLUTION

1. a–c. Complete answers should address these points:

a. • Explain that while both graphs are increasing from left to right, the graph of $g(x)$ has $y < 0$ for all values of x. The graph of $f(x)$ has $y > 0$ for all values of x.

b. • Explain that $h(x)$ and $j(x)$ are both decreasing from left to right although $j(x)$ has $y > 0$ for all values of x, and $h(x)$ has $y < 0$ for all values of x. The logarithmic graph is increasing. It begins with negative y-values, and increases to positive y-values.

• Explain that the exponential graphs have a horizontal asymptote, and the logarithmic graph has a vertical asymptote.

c. • $y = a \cdot 3^x$ and $y = \log_3 x$; Their graphs are reflections of each other in the line $y = x$. It can also be determined by rewriting the log equation in exponential form.

MULTI-STEP PROBLEM SOLUTION

2. a. $P(t) = 50{,}000(1.051)^t$; 74,437

b. $P(t) = 100{,}000(0.919)^t$

c. 12,103; *Sample Answer:* The model began in 1980 and was good for 10 years. To estimate 25 years out is probably not accurate.

d. *Sample Answer:* Town A has exponential growth, Town B is exponential decay. Exponential models can be used because there is a constant percent increase or decrease each year.

e. about 14 years

f. about 3.4 years

3. *Sample Answer:* 1985; One method would be to solve algebraically, another to solve graphically, or to solve using a table of values. Algebraically is more challenging since there are exponents with different bases on both sides.

MULTI-STEP PROBLEM RUBRIC

4 Students complete all parts of the questions accurately. Explanations are clear and logical, showing an understanding of exponential growth, exponential decay, and logarithms. Students demonstrate a thorough understanding of both algebraic and graphical solutions.

3 Students complete the questions and explanations. Explanations may be somewhat vague, but show some understanding of exponential growth, exponential decay, and logarithms. Students demonstrate some understanding of both algebraic and graphical solutions.

2 Students complete questions and explanations. Explanations are not correct. Students are unable to show understanding of exponential growth, exponential decay, and logarithms. Examples are incomplete or inaccurate.

1 Students' work is very incomplete. Solutions or reasoning are incorrect. Examples are incomplete or inaccurate.

NAME ______________________ DATE ____________

Project: Investigating Zipf's Law

For use with Chapter 8

OBJECTIVE **Use Zipf's Law to model the relationship between population and rank for the cities in a country.**

MATERIALS Encyclopedia, almanac, or Internet access; graphing calculator

INVESTIGATION In Exercise 57 of Section 8.7, you investigated the relationship between population and rank for cities in Argentina. Zipf's Law makes a generalization about the relationship between these two variables.

> **ZIPF'S LAW**
>
> For the cities in any country, rank is a power function of population.

1. Choose a country that is of interest to you.
2. Use your encyclopedia, almanac, or the Internet to make a list of the country's largest cities, their populations, and their ranks. Your list should contain at least ten cities. (If you have trouble finding data for at least ten cities, choose another country.)
3. Make a scatter plot of the data pairs $(\ln P, \ln R)$ for your cities, where P is the population and R is the rank. Fit a line to the plotted points, and find an equation of the line.
4. Is a power model a good fit for the data?
5. Find a power model for your data.
6. Use your power model to predict the rank of some city not used in your scatter plot. How well does the predicted rank match the actual rank?
7. If you have enough time, repeat this investigation for one or two additional countries. You can also try using states instead of countries.

PRESENT YOUR RESULTS Write a report summarizing your results. Your report should include a statement of the goal of your project, a brief description of your chosen country and its people, a map, and your answers to the numbered questions above. How well do you think Zipf's Law describes the relationship between rank and population for various countries?

Project: Teacher's Notes

For use with Chapter 8

GOALS
- Evaluate logarithms.
- Graph a set of data and find a power model.
- Analyze results.

MANAGING THE PROJECT Students can work in groups of 2 to 4 students.

Since it may be difficult to use the Internet to find the data for foreign countries, you may want to suggest that students start by using United States data. A good source for United States data is www.census.gov.

RUBRIC The following rubric can be used to assess student work.

4 The student locates appropriate data and produces an accurate scatter plot. The student correctly determines a power model and calculates the predicted rank for a city not on his or her original list. The report presents results accurately, neatly, and completely. The student reaches an appropriate conclusion regarding how well Zipf's Law describes the relationship between rank and population.

3 The student locates appropriate data and produces a reasonably accurate scatter plot. The student attempts to determine a power model and calculate the predicted rank for a city not on his or her original list. However, the student may not perform all calculations accurately or may make a minor conceptual error in applying logarithms. The report may contain minor errors, be somewhat sloppy, or omit minor details.

2 The student attempts to locate appropriate data, graph the data, and determine a power model. However, work may be incomplete or reflect misunderstandings. For example, the student may plot $\ln R$ versus P instead of versus $\ln P$. The report may indicate a limited grasp of certain ideas or may lack key elements.

1 Equations, graphs, interpretations of graphs, and predictions using a power model are missing or do not show an understanding of key ideas. The student doesn't give a reasonable conclusion regarding Zipf's Law or fails to support the conclusion.

NAME ______________________________ DATE __________

Cumulative Review

For use after Chapters 1–8

Solve the equation. (1.3)

1. $3a + 5 = 7a - 8$
2. $x - 5 + 4 = 3(2 - x)$
3. $-10b + 5 = 5b$
4. $\frac{2}{3}(a - 2) + 6 = 18$
5. $\frac{1}{4}x + \frac{1}{5} = \frac{3}{5}x - \frac{1}{8}$
6. $x + 0.05 = 2.3$

Use slope-intercept form to graph the equation. (2.3)

7. $y = 3x + 1$
8. $y = \frac{1}{3}x - 2$
9. $y = 4x + \frac{3}{4}$
10. $4x + 2y = 8$
11. $3x - 6y = 18$
12. $y = -\frac{5}{2}x - \frac{2}{3}$

Write an equation of the line from the given information. (2.3, 2.4)

13. The line passes through $(2, 1)$ and $(7, -8)$.
14. The line has a slope of $\frac{2}{3}$ and a y-intercept of -60.
15. The line passes through $(1, -2)$ and is perpendicular to the line $y = \frac{4}{3}x - 3$.
16. The line passes through $(3, 4)$ and is parallel to the line that passes through $(3, 8)$ and $(5, 10)$.
17. The line passes through $(2, -6)$ and is parallel to $x = 8$.
18. The line passes through $(3, -5)$ and is perpendicular to $x = 10$.

Graph the linear system and estimate the solution. Then check the solution algebraically. (3.1)

19. $4x + 2y = 14$
    $3x - 5y = -22$
20. $x - 3y = 5$
    $-2x + 2y = -6$
21. $5x + 2y = 10$
    $-4x - 3y = -15$
22. $3x - 2y = 12$
    $2x + y = 1$
23. $-3x + 5y = -7$
    $-2x + y = -7$
24. $5x + 3y = 10$
    $-4x + 8y = -8$

Use an inverse matrix to solve the linear system. (4.5)

25. $4x - 2y = 10$
    $3x - y = 7$
26. $x + y = -8$
    $2x - 8y = 14$
27. $4x - 2y = 8$
    $8x + 4y = 16$
28. $2x - 3y = 27$
    $3x - y = 23$
29. $-x + 5y = -2$
    $-2x + 6y = -12$
30. $4x + 9y = 5$
    $6x + 6y = 5$

Write the quadratic function in standard form. (5.1)

31. $y = (x - 3)(x - 4)$
32. $y = -(x + 5)(2x + 3)$
33. $y = -2(x + 3)^2 - 1$
34. $y = -(3x - 2)^2 + 6x$
35. $y = 5(x - 2)^2 + 2$
36. $y = \frac{1}{2}(2x - 3)^2 + \frac{1}{2}$

Solve the quadratic equation. (5.2)

37. $x^2 - 5x + 4 = 0$
38. $9a^2 + 12a + 4 = 0$
39. $30x^2 - 60x = 0$
40. $-a^2 - 19a - 88 = 0$
41. $5a^2 - 16 = 4a^2 - 7$
42 $3x^2 - 5x + 7 = 2x^2 - 2x + 11$

Simplify the expression. (5.3)

43. $\sqrt{27}$
44. $2\sqrt{50}$
45. $\sqrt{5} \cdot \sqrt{10}$

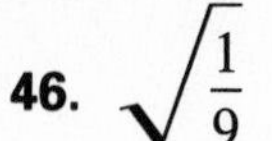

CHAPTER 8 CONTINUED

NAME ______________________ DATE ________

Cumulative Review

For use after Chapters 1–8

Simplify the expression. (5.3)

46. $\sqrt{\frac{1}{9}}$ **47.** $\frac{2}{\sqrt{5}}$ **48.** $\frac{\sqrt{7}}{\sqrt{32}}$

Write the expression as a complex number in standard form. (5.4)

49. $i(4 + 3i)$ **50.** $(6 + i)(7 - 3i)$ **51.** $(3 + 2i)^2$

52. $\frac{8}{2 + i}$ **53.** $\frac{-4 - 3i}{2i}$ **54.** $\frac{7 + 3i}{2 + 5i}$

Solve the equation by completing the square. (5.5)

55. $x^2 + 4x = -2$ **56.** $x^2 - 5x + 7 = 0$ **57.** $u^2 - 2u = 4u + 8$

58. $2x^2 + 4x = 1$ **59.** $-3x^2 + 6x + 2 = 0$ **60.** $4r^2 - 9r = -r + 5$

Find the discriminant of the quadratic equation and give the number and type of solutions of the equation. (5.6)

61. $x^2 + 4x + 4 = 0$ **62.** $2x^2 - 5x + 3 = 0$ **63.** $3x^2 - x + 2 = 0$

64. $4x^2 - 10 = 0$ **65.** $64x^2 + 16x + 1 = 0$ **66.** $4x^2 - 9x = 0$

Evaluate the expression. (6.1)

67. $5^{-4} \cdot 5^2$ **68.** $\frac{6^2}{6^5}$ **69.** $(3^{-3})^{-2}$

70. $\left(\frac{1}{3}\right)^{-3}$ **71.** $9^{-3} \cdot 9^0$ **72.** $\left(\frac{2}{3}\right)^{-2}$

Divide using polynomial long division. (6.5)

73. $(x^2 + 6x + 8) \div (x + 4)$ **74.** $(2x^2 + 3x - 2) \div (2x - 1)$

75. $(6x^2 + x + 3) \div (3x + 2)$ **76.** $(2x^3 + x^2 - 7x + 7) \div (2x - 1)$

77. $(x^3 + 3x^2 + 6x + 8) \div (x^2 + 3)$ **78.** $(x^4 + 2x^3 + 5x^2 + 10x + 5) \div (x^2 + 5)$

Solve the equation. Check for extraneous solutions. (7.6)

79. $x^{3/2} = 125$ **80.** $x^{1/5} - 2 = 0$ **81.** $2(3x - 1)^{1/2} = 8$

82. $\sqrt{x - 25} = 10$ **83.** $x - 8 = \sqrt{x - 2}$ **84.** $x + 5 = \sqrt{20x + 9}$

85. ***Land Value*** You purchased land for \$50,000 in 1980. The value of the land increased by approximately 4% per year. What is the approximate value of the land in the year 2000? **(8.1)**

86. ***Depreciation*** You buy a new car for \$21,000. It depreciates by 10.5% each year. Estimate when the car will have a value of \$17,000. **(8.2)**

87. ***Continuous Compounding*** You deposit \$850 in an account that pays 6.5% annual interest compounded continuously. What is the balance after 5 years? **(8.3)**

Review and Assess

ANSWERS

Chapter Support

Parent Guide

8.1: $p = 10{,}000(1.2)^t$; about 8 minutes
8.2: $h = 5(0.65)^n$; 4 bounces **8.3:** about 8.166
8.4: $3x$ 8.5: $\log_7\left(\frac{x^5}{2}\right)$ **8.6:** 3 years
8.7: $y = 22{,}500(0.8)^x$; \$9216 **8.8:** 3

Prerequisite Skills Review

1. $\frac{1}{27}$ **2.** $\frac{9}{4}$ **3.** $-\frac{1}{49}$ **4.** $\frac{9}{49}$

5. $f(x) \to +\infty$ as $x \to -\infty$ and $f(x) \to +\infty$ as $x \to +\infty$ **6.** $f(x) \to +\infty$ as $x \to -\infty$ and $f(x) \to -\infty$ as $x \to +\infty$ **7.** $f(x) \to -\infty$ as $x \to -\infty$ and $f(x) \to -\infty$ as $x \to +\infty$

8. $f(x) \to -\infty$ as $x \to -\infty$ and $f(x) \to +\infty$ as $x \to +\infty$ **9.** *Sample answer:* $y = 0.45x + 1.7$

Strategies for Reading Mathematics

1. 5; $5^3 = 125$ **2. a.** $\log_2\left(\frac{1}{4}\right) = -2$
b. $\log_6 216 = 3$ **c.** $\log_9 81 = 2$
d. $\log_{16} 4 = \frac{1}{2}$ **3. a.** $10^{-3} = 0.001$
b. $7^2 = 49$ **c.** $2^5 = 32$ **d.** $8^{1/3} = 2$ **4. a.** 2
b. -2 **c.** 5 **d.** $\frac{1}{2}$

Lesson 8.1

Warm-up Exercises

1. $f(x) \to \infty$ **2.** $f(x) \to -\infty$ **3.** $f(x) \to \infty$
4. domain: all real numbers; range: $y \ddagger 0$
5. domain: $x \ddagger 0$; range $y \ddagger 0$

Daily Homework Quiz

1. \$1.16, \$1.19, \$1.19 **2.** \$1.20, \$0.32
3.

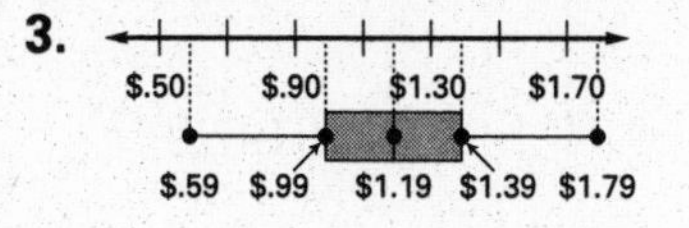

Lesson Opener

Allow 15 minutes.

1. *Sample answer:* No, the king should have done some calculations to make sure that the request was reasonable. **2.** Number of grains: 8, 16, 32 Exponential form: 2^3, 2^4, 2^5 **3.** 2^{x-1}

4. $2^{63} \approx 9.22 \times 10^{18}$ **5.** about 738 million pounds per person **6.** no

Practice A

1. F **2.** B **3.** A **4.** E **5.** C **6.** D

7. Shift graph of f 2 units up. **8.** Shift graph of f 5 units down. **9.** Shift graph of f 1 unit left. **10.** Shift graph of f 3 units right. **11.** Reflect graph of f across x-axis. **12.** Shift graph of f 2 units up. **13.** 1; x-axis **14.** 1; x-axis **15.** 2; x-axis **16.** $\frac{1}{2}$; x-axis **17.** -2; x-axis **18.** $-\frac{1}{4}$; x-axis **19. a.** \$2100 **b.** \$2101.89 **c.** \$2102.32

Practice B

1. B **2.** A **3.** C **4.** E **5.** F **6.** D

7. Shift the graph of f 1 unit right and 2 units up.

8. Shift the graph of f 2 units left and reflect across the x-axis. **9.** Shift the graph of f 2 units left and 4 units down. **10.** 3; $y = 2$
11. $\frac{1}{27}$; x-axis **12.** 1; $y = -2$

13.

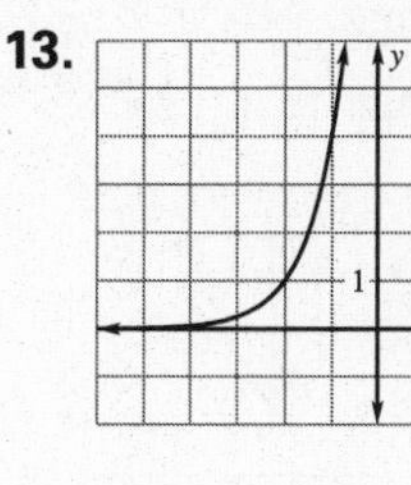

14.

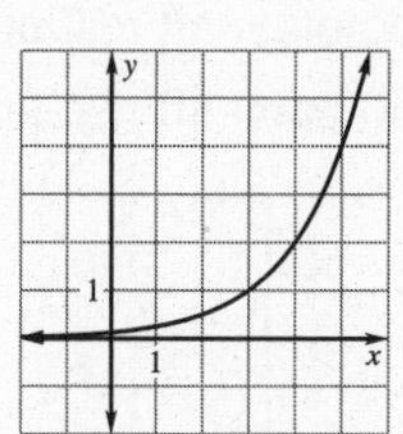

15.

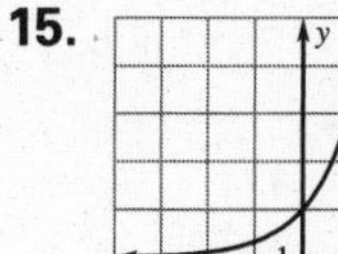

16.

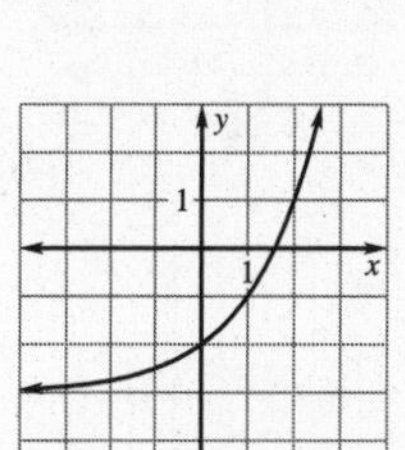

Lesson 8.1 *continued*

17. 18.

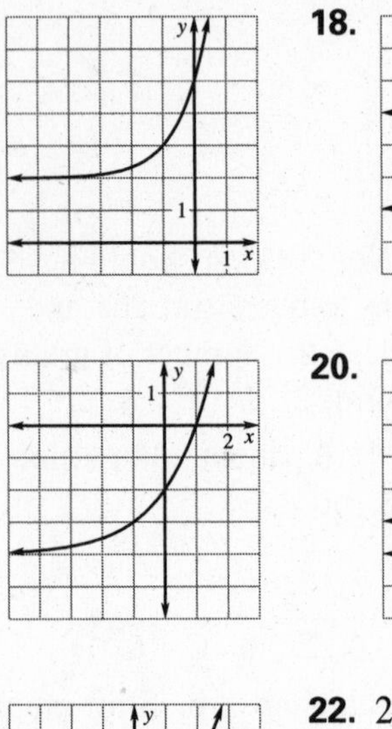

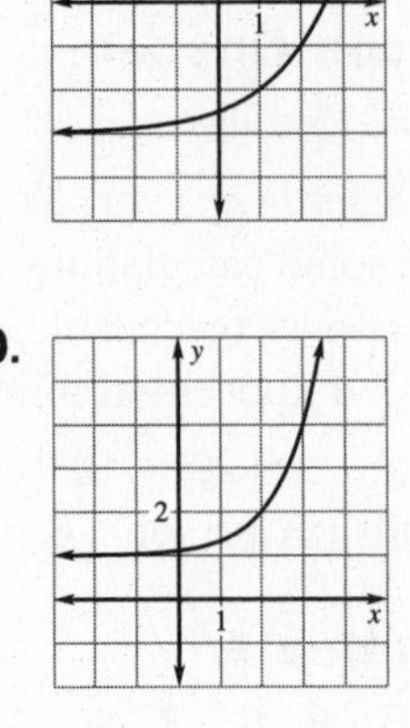

19. 20.

21.

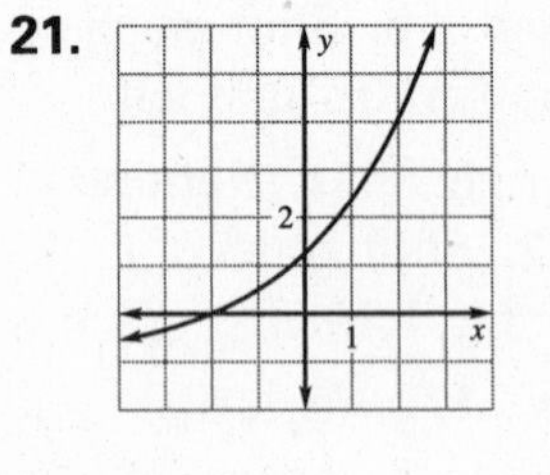

22. 25.2; 1.15; 15%

23.

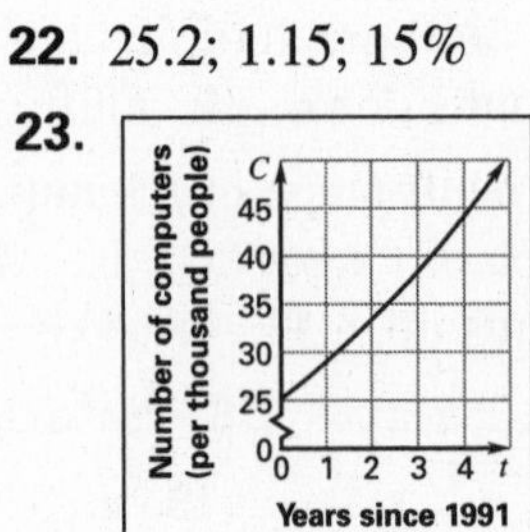

24. 88.7

Practice C

1. -1; $y = -2$ 2. $\frac{1}{5}$; x-axis 3. 7; $y = 4$
4. $\frac{1}{75}$; x-axis 5. 6; $y = 7$ 6. -3.2×10^{-5}; x-axis 7. domain: all real numbers; range: $y > 3$
8. domain: all real numbers; range: $y > -2$
9. domain: all real numbers; range: $y > 4$
10. domain: all real numbers; range: $y > -2$
11. domain: all real numbers; range: $y > 4$
12. domain: all real numbers; range: $y < -3$

13. 14.

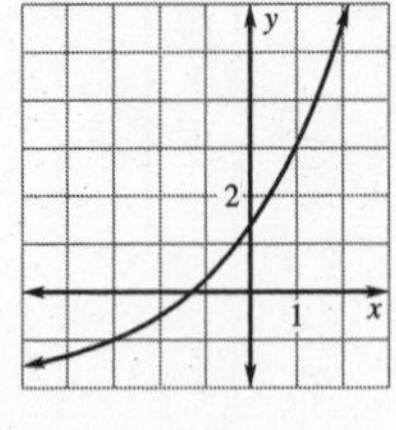

15. 16.

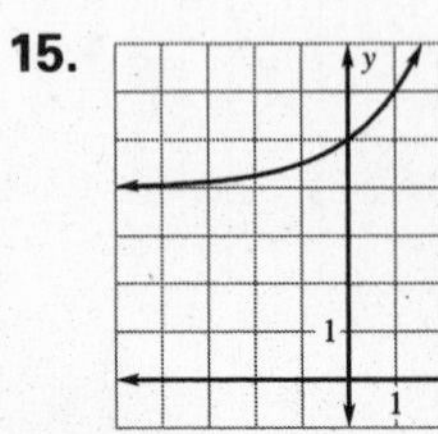

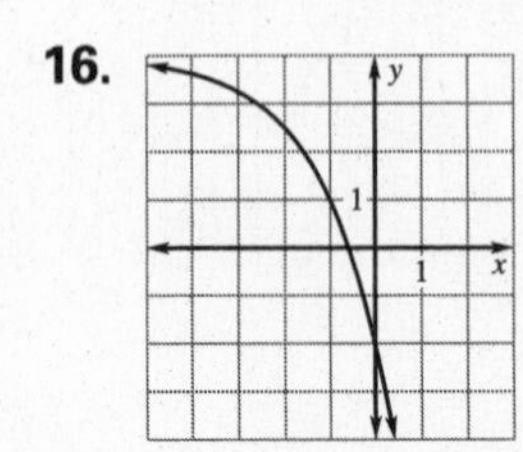

17. 18.

19. 20.

21.

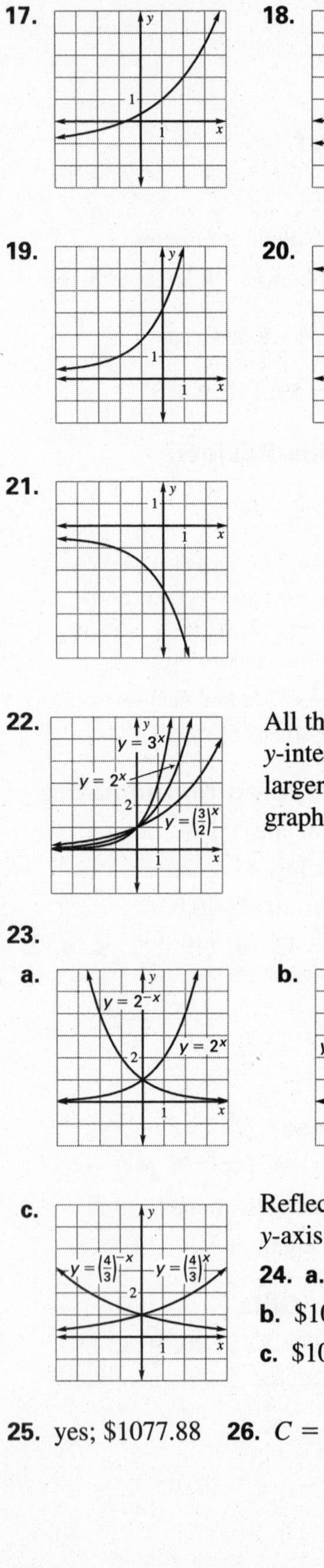

22. All three graphs have a y-intercept of 1. The larger a is, the steeper the graph.

23.

a.

b.

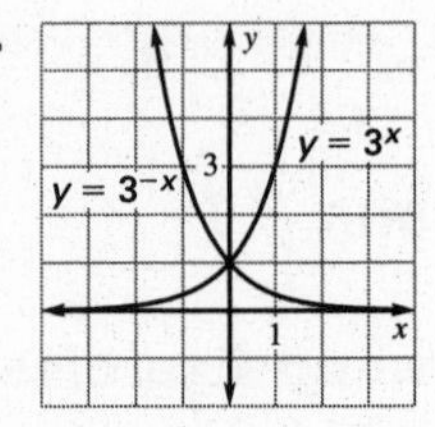

c. Reflection across the y-axis

24. a. \$1077.80

b. \$1077.88

c. \$1077.88

25. yes; \$1077.88 26. $C = 15{,}000(1.072)^t$

27.

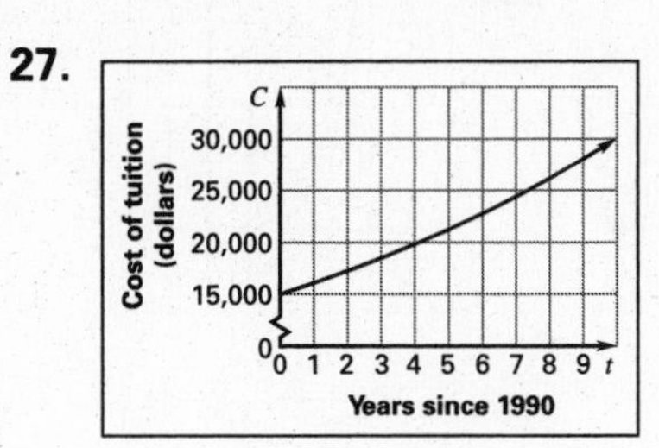

28. 1994 **29.** $60,254.15

Reteaching with Practice

1.

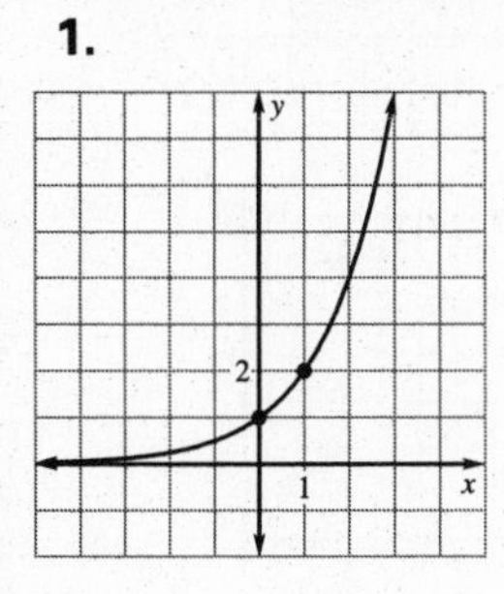

2.

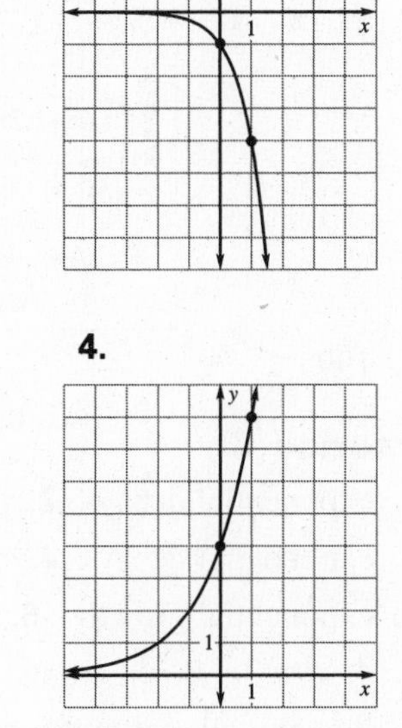

3.

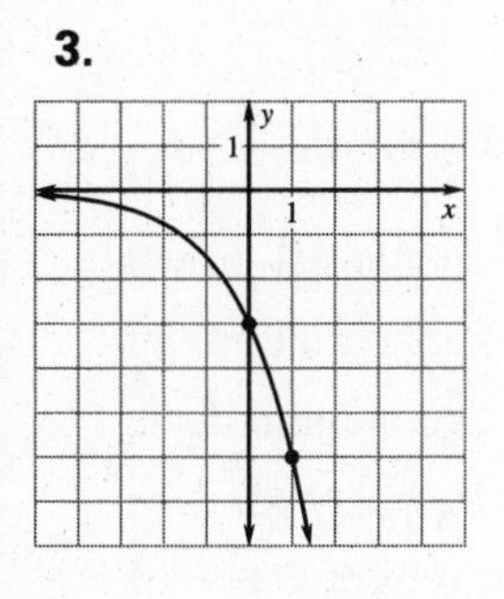

4.

5. domain: all real numbers; range: $y < 0$

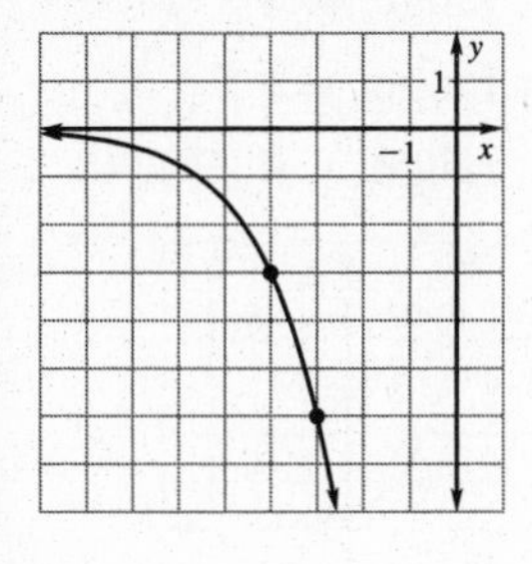

6. domain: all real numbers; range: $y > 0$

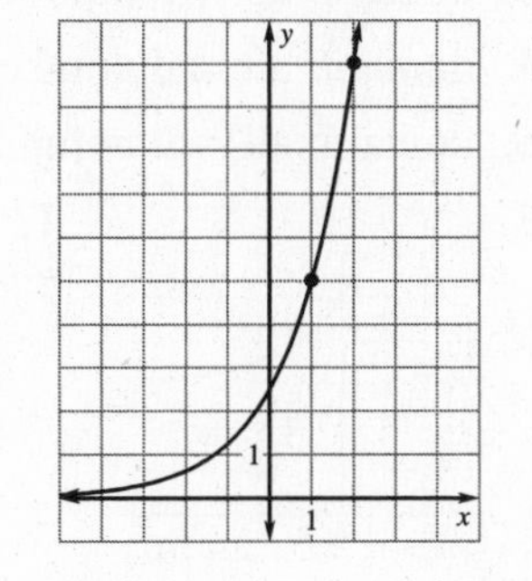

7. domain: all real numbers; range: $y > 4$ **8.** domain: all real numbers; range: $y > -3$

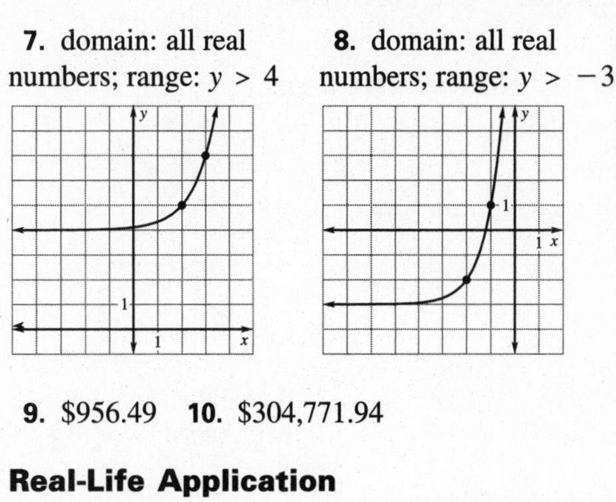

9. $956.49 **10.** $304,771.94

Real-Life Application

1. 1.413; 41.3%

2.

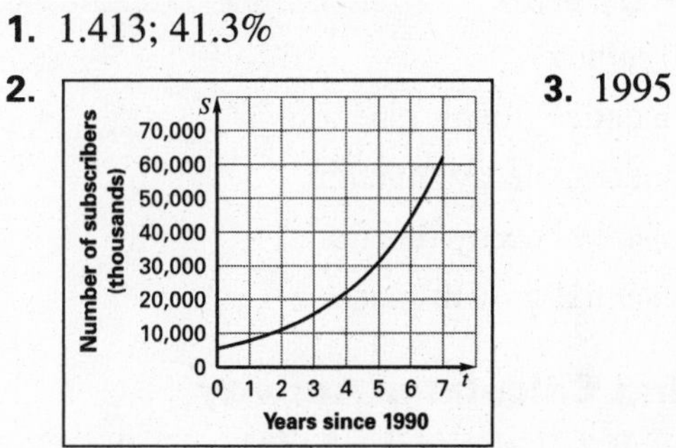

3. 1995

4. 1998 **5.** 350.6 million; 989.2 million; 5.6 billion **6.** No, the numbers begin to increase much too rapidly.

Challenge: Skills and Applications

1. $y = 4 \cdot 4^x$ **2.** $y = \frac{5}{3} \cdot 3^x$

3. $y = \sqrt{5} \cdot \left(2\sqrt{2}\right)^x$ **4.** $y = \frac{\sqrt{2}}{3}\left(\frac{1}{25}\right)^x$

5. a. $129.52; $1042.70 **b.** yes **6. a.** 1, 2, 4, 8, 16; 2^x **b.** 2, 6, 18, 54, 162, $2 \cdot 3^x$ **c.** 4, 20, 100, 500, 2500, $4 \cdot 5^x$ **d.** $(a - 1)a^x$

7. average growth rate $= \frac{3^{a+h} - 3^a}{h} =$

$3^a \cdot \frac{3^h - 1}{h}$; 1.10

Lesson 8.2

Warm-up Exercises

1. $\frac{1}{2}$ **2.** 3.5 **3.** -2

4. domain: all real numbers; range: $y > 0$

5. domain: all real numbers; range: $y < 0$

Lesson 8.2 *continued*

Daily Homework Quiz

1. 4.5; the x-axis
2.
3. 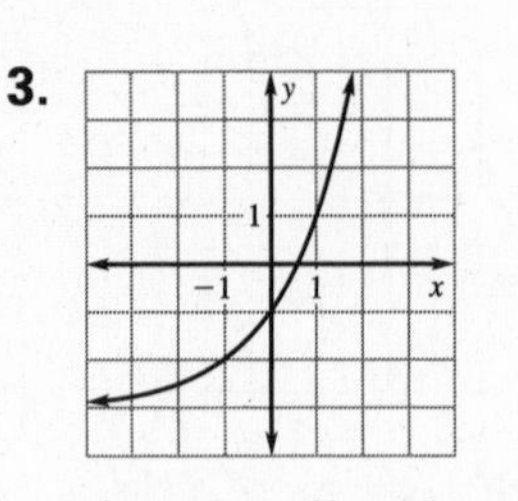

domain: all real numbers; range: $y > -3$

Lesson Opener

Allow 10 minutes.

1. exponential growth function
2. exponential decay function
3. exponential decay function
4. exponential growth function

Graphing Calculator Activity

1. rise, fall, fall, rise, rise **3.** If $b > 1$, the graph will rise; if $0 < b < 1$, the graph will fall.
4. a. fall **b.** fall **c.** rise **d.** fall **e.** rise **f.** rise

Practice A

1. exponential decay **2.** exponential growth
3. exponential growth **4.** exponential decay
5. exponential decay **6.** exponential growth
7. A **8.** E **9.** D **10.** F **11.** C **12.** B
13. 1; x-axis **14.** 1; x-axis **15.** 2; x-axis
16. $\frac{1}{4}$; x-axis **17.** -5; x-axis **18.** $-\frac{2}{3}$; x-axis
19. 8.78 grams

Practice B

1. exponential decay **2.** exponential growth
3. exponential decay **4.** E **5.** A **6.** C
7. F **8.** D **9.** B
10. **11.** 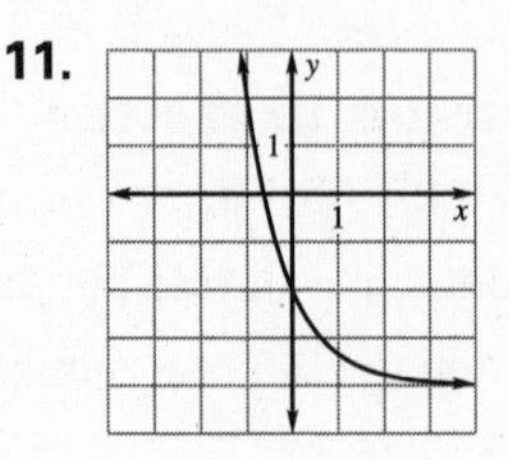

12. **13.** **14.** **15.** **16.** $1.14 **17.**

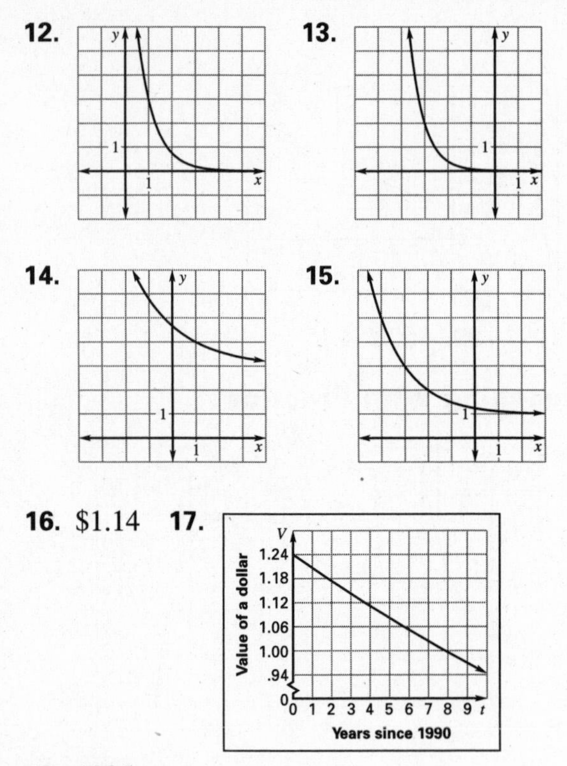

18. 1995

Practice C

1. exponential decay **2.** exponential growth
3. exponential decay **4.** exponential growth
5. exponential growth **6.** exponential decay
7. 4; $y = 3$ **8.** $\frac{8}{27}$; x-axis **9.** $\frac{3}{16}$; x-axis
10. domain: all real numbers; range: $y > -3$
11. domain: all real numbers; range: $y > 4$
12. domain: all real numbers; range: $y > 1$
13. domain: all real numbers; range: $y > -2$
14. domain: all real numbers; range: $y > 7$
15. domain: all real numbers; range: $y < -4$

16. 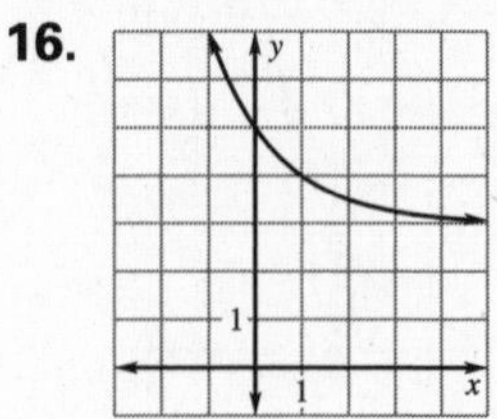
17. 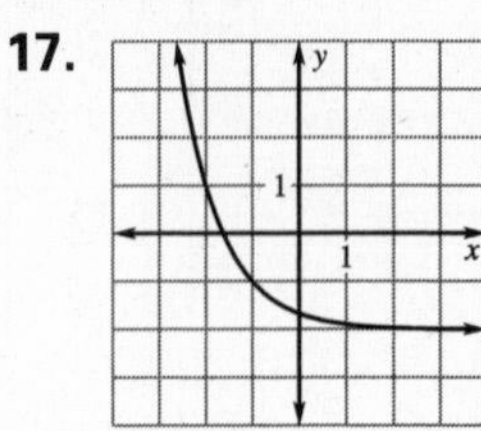
18. 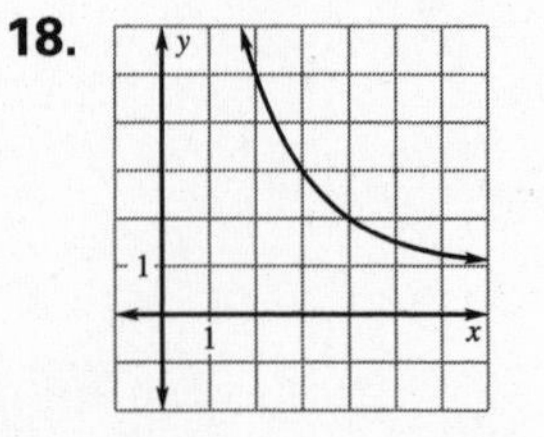
19. 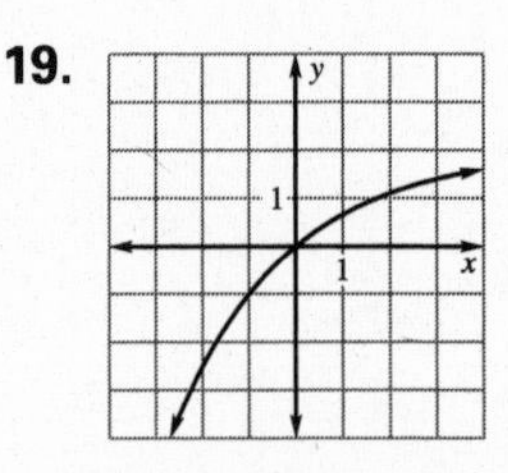

Lesson 8.2 *continued*

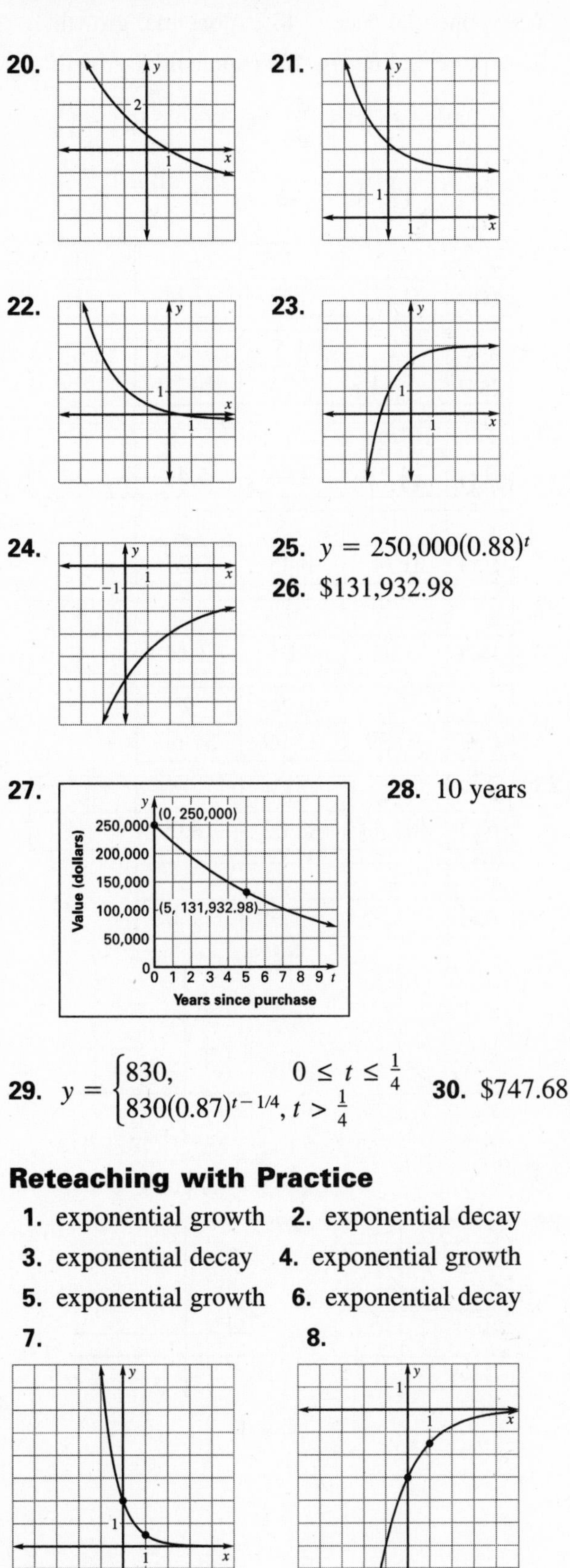

20. 21. 22. 23. 24.

25. $y = 250{,}000(0.88)^t$

26. \$131,932.98

27.

28. 10 years

29. $y = \begin{cases} 830, & 0 \le t \le \frac{1}{4} \\ 830(0.87)^{t - 1/4}, & t > \frac{1}{4} \end{cases}$

30. \$747.68

Reteaching with Practice

1. exponential growth 2. exponential decay
3. exponential decay 4. exponential growth
5. exponential growth 6. exponential decay
7. 8.

9.

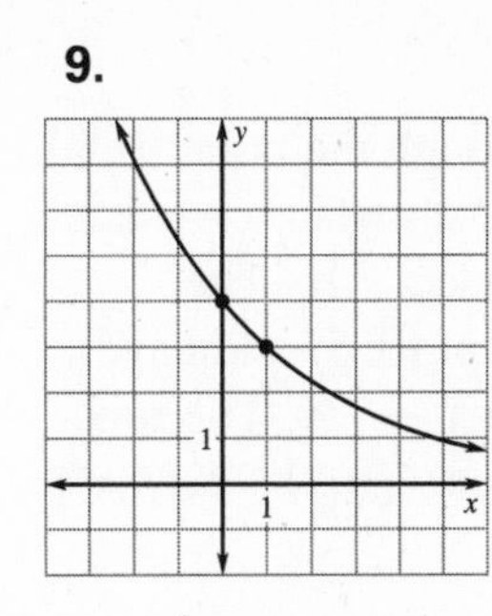

10.

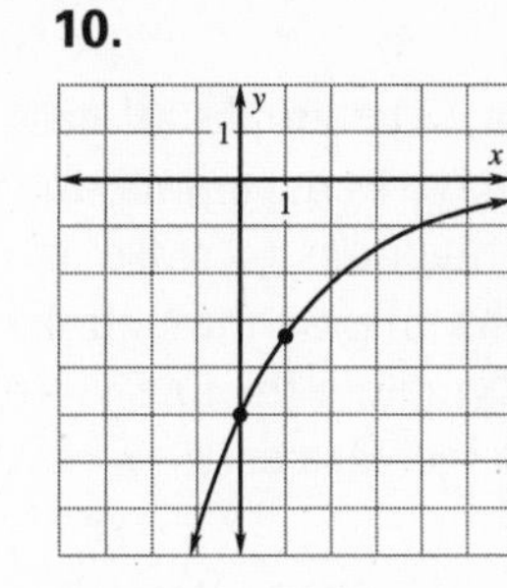

11. domain: all real numbers; range: $y > 0$

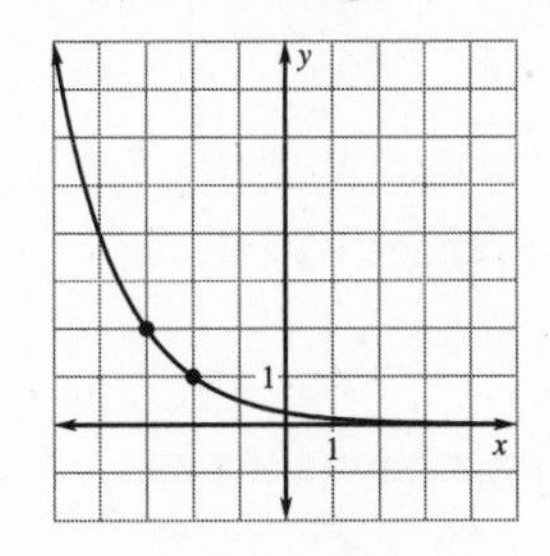

12. domain: all real numbers; range: $y < 0$

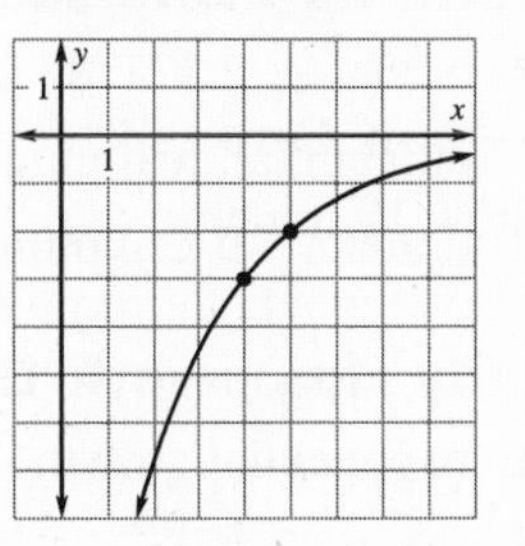

13. domain: all real numbers; range: $y < 2$

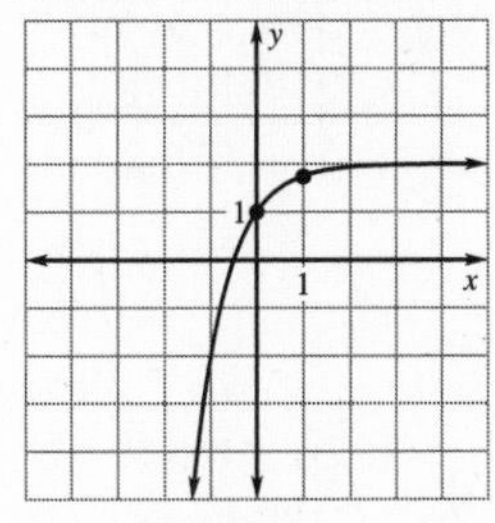

14. domain: all real numbers; range: $y > -3$

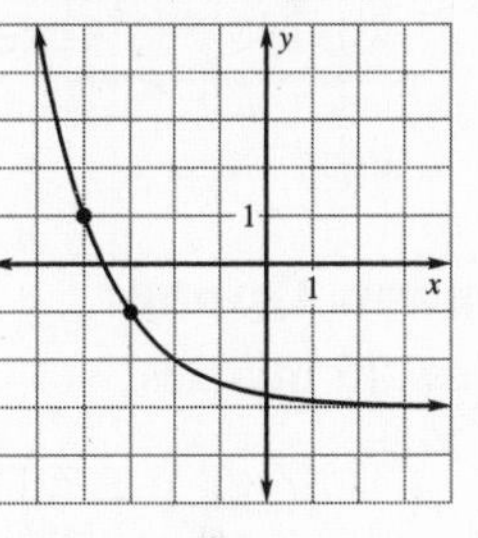

Interdisciplinary Application

1. 28.5; 0.9567; 4.33%

2.

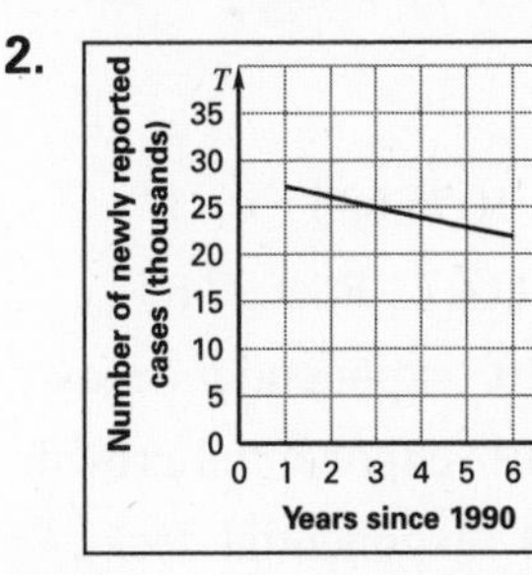

3. 1993 4. 2003

5. about 14,700

Challenge: Skills and Applications

1. $\frac{8}{3}, \frac{1}{6}$ 2. $\frac{32}{81}$; $\left(a = 3, b = \frac{2}{3}\right)$ 3. a. $V = V_0p^t$ b. \$3797 4. a. $N = N_o\left(\frac{1}{2}\right)^{t/5700}$ b. 17,100 yr

Lesson 8.2 *continued*

5. a. The graphs all pass through the point (0, 3) and have as asymptote the line $y = 0$ as $x \to \infty$. The functions increase as $x \to \infty$. **b.** The graphs all pass through the point (0, 5) and have as asymptote the line $y = 0$ as $x \to \infty$. The functions increase as $x \to \infty$. **6. a.** $f(nx)$ **b.** $f(x^n)$

Lesson 8.3

Warm-up Exercises

1. 2.25 **2.** 2.37 **3.** 2.44 **4.** 2.49

5. domain: all real numbers; range: $y > 0$

6. domain: all real numbers; range: $y < 0$

Daily Homework Quiz

1. exponential growth **2.** exponential decay

3.

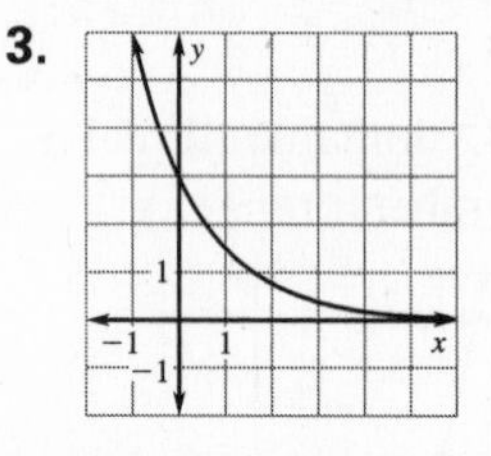

4.

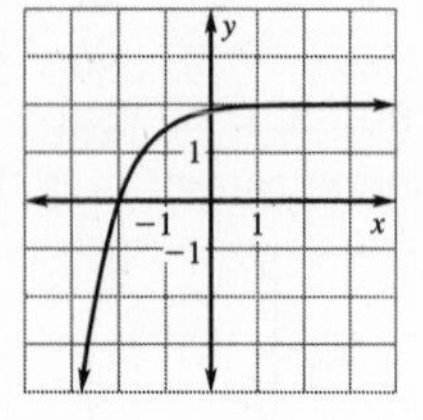

Lesson Opener

Allow 15 minutes.

1. $A = 1000\left(1 + \frac{1}{n}\right)^n$

2. 2000.00, 2441.41, 2613.04, 2714.57, 2718.13, 2718.28 **3.** \$2,718.28

Practice A

1. 54.598 **2.** 0.368 **3.** 1096.633 **4.** 1

5. 0.135 **6.** 1.948 **7.** 0.607 **8.** 9.974

9. exponential growth **10.** exponential decay

11. exponential growth **12.** exponential growth

13. exponential decay **14.** exponential decay

15. e^8 **16.** e^6 **17.** e^{10} **18.** e^3 **19.** $e^{-5} = \frac{1}{e^5}$

20. $8e^{15}$ **21.** A **22.** C **23.** B **24.** \$829.79

25. 273,544

Practice B

1. 148.413 **2.** 0.717 **3.** 0.247 **4.** 4.113

5. exponential growth **6.** exponential decay

7. exponential decay **8.** exponential growth

9. exponential decay **10.** exponential growth

11. $\frac{1}{e^8}$ **12.** $3e^4$ **13.** $\frac{2}{e}$ **14.** $16e^6$ **15.** $-12e^3$

16. $2e^{2x+3}$ **17.** $8e^{2x}$ **18.** e **19.** $\frac{1}{e^x}$

20.

x	−2	−1.5	−1	0
$f(x)$	0.27	0.45	0.74	2

x	1	1.5	2
$f(x)$	5.44	8.96	14.78

21.

x	−2	−1.5	−1	0
$f(x)$	14.78	8.96	5.44	2

x	1	1.5	2
$f(x)$	0.74	0.45	0.27

22.

x	−2	−1.5	−1	0
$f(x)$	3.02	3.05	3.14	4

x	1	1.5	2
$f(x)$	10.39	23.09	57.60

23.

x	−2	−1.5	−1	0
$f(x)$	401.43	88.02	18.09	−1

x	1	1.5	2
$f(x)$	−1.95	−1.99	−2.00

24.

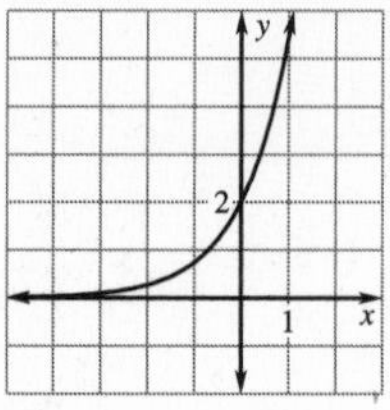

$y = 0$

25.

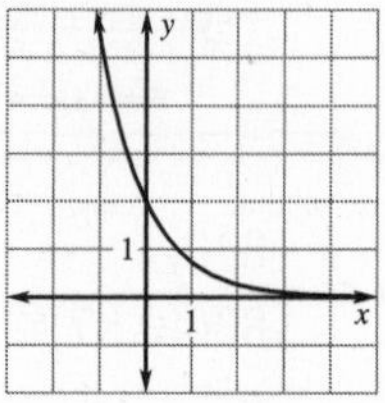

$y = 0$

26.

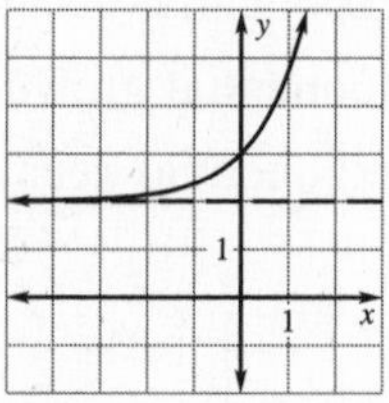

$y = 2$

27.

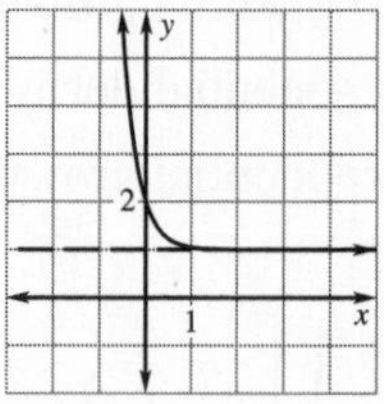

$y = 1$

Lesson 8.3 *continued*

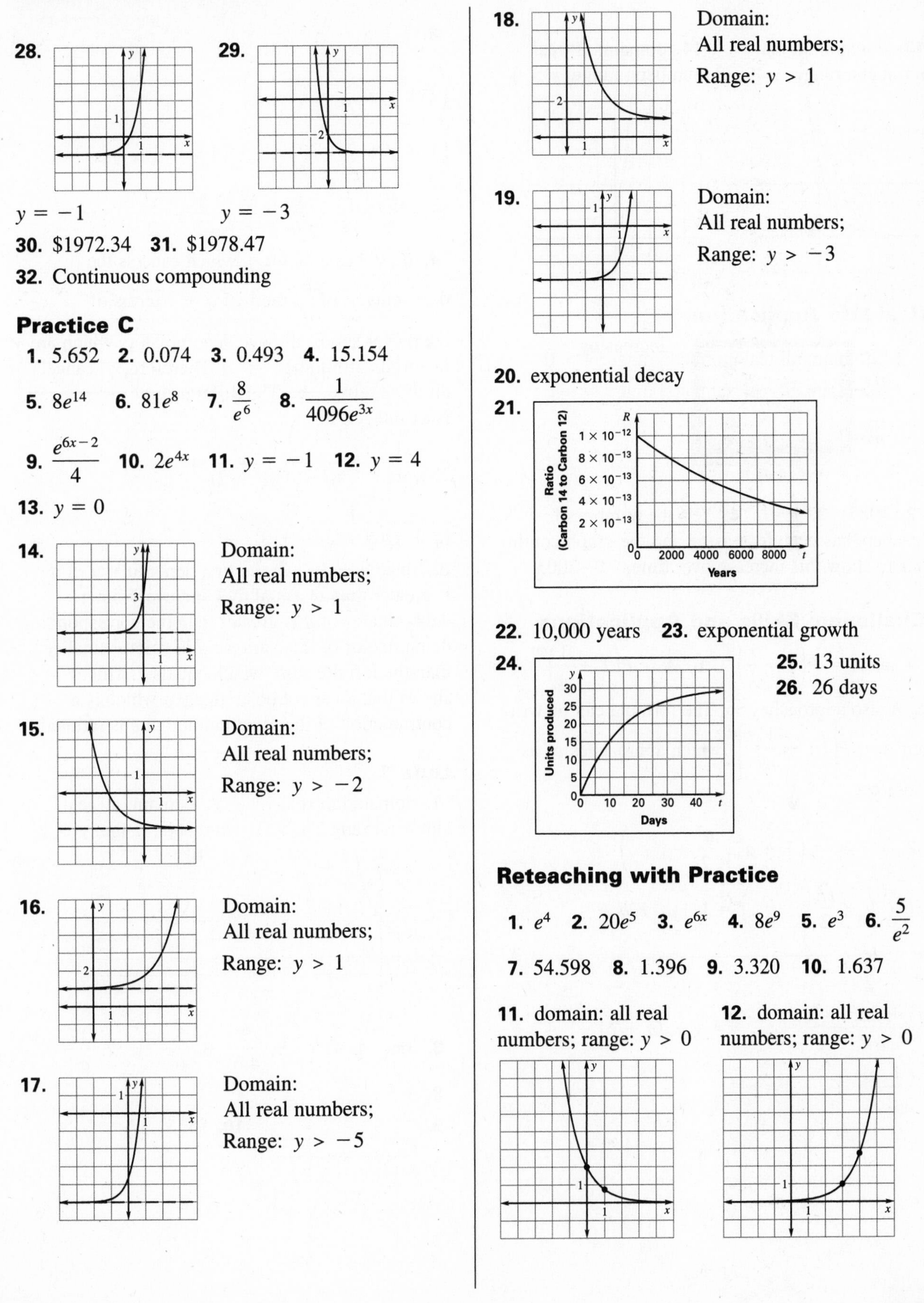

28. $y = -1$

29. $y = -3$

30. \$1972.34 **31.** \$1978.47

32. Continuous compounding

Practice C

1. 5.652 **2.** 0.074 **3.** 0.493 **4.** 15.154

5. $8e^{14}$ **6.** $81e^{8}$ **7.** $\frac{8}{e^6}$ **8.** $\frac{1}{4096e^{3x}}$

9. $\frac{e^{6x-2}}{4}$ **10.** $2e^{4x}$ **11.** $y = -1$ **12.** $y = 4$

13. $y = 0$

14. Domain: All real numbers; Range: $y > 1$

15. Domain: All real numbers; Range: $y > -2$

16. Domain: All real numbers; Range: $y > 1$

17. Domain: All real numbers; Range: $y > -5$

18. Domain: All real numbers; Range: $y > 1$

19. Domain: All real numbers; Range: $y > -3$

20. exponential decay

21.

22. 10,000 years **23.** exponential growth

24.

25. 13 units

26. 26 days

Reteaching with Practice

1. e^4 **2.** $20e^5$ **3.** e^{6x} **4.** $8e^9$ **5.** e^3 **6.** $\frac{5}{e^2}$

7. 54.598 **8.** 1.396 **9.** 3.320 **10.** 1.637

11. domain: all real numbers; range: $y > 0$

12. domain: all real numbers; range: $y > 0$

Lesson 8.3 *continued*

13. domain: all real numbers; range: $y > -3$

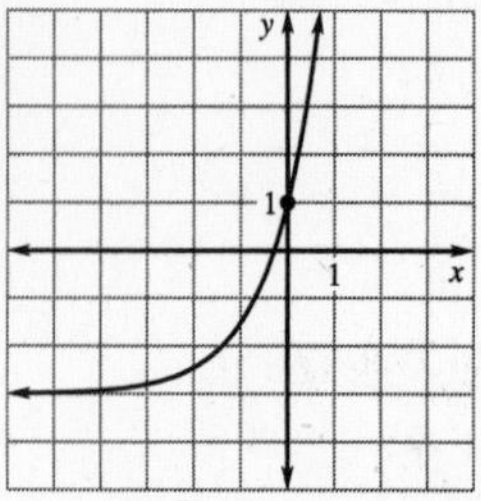

14. domain: all real numbers; range: $y > 1$

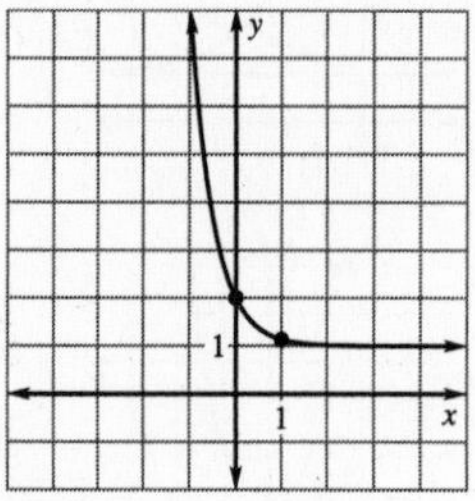

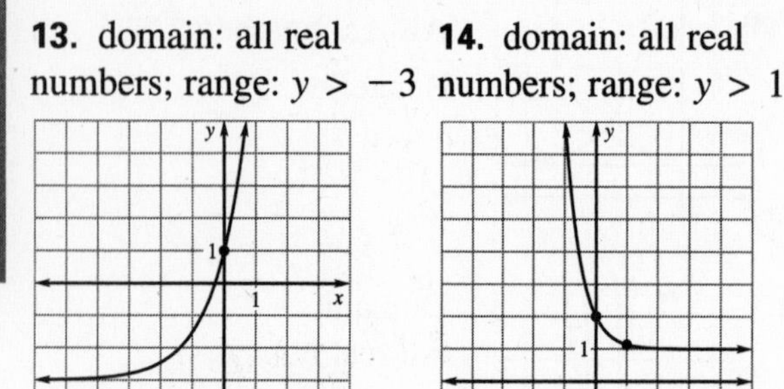

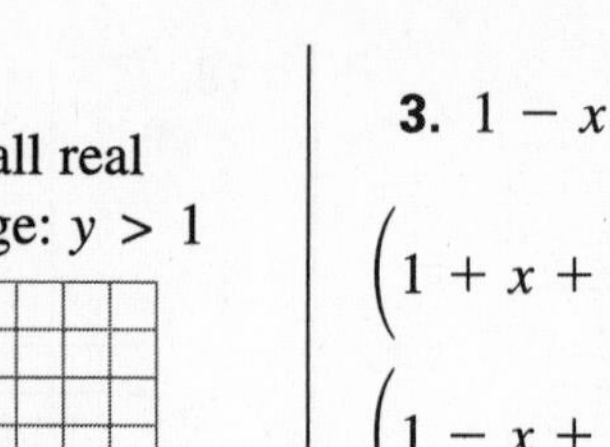

Real-Life Application

1.

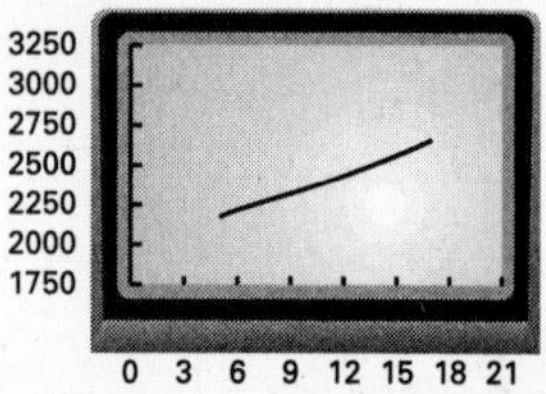

increasing

2. 1987 **3.** 1991 **4.** Yes, the number of teachers has been increasing and the graph continues to show this increase over time. **5.** 2003

Challenge: Skills and Applications

1. a. $A = P\left(1 + \frac{r}{k}\right)^{kt}$ **b.** $A = P\left(1 + \frac{1}{n}\right)^{nrt}$

c. n also approaches ∞. The formula can be written $A = P\left[\left(1 + \frac{1}{n}\right)^{n}\right]^{rt}$ which approaches e^{rt} as $k \to \infty$.

2. $e^a \cdot e^b = \left(1 + a + \frac{a^2}{2} + \ldots\right)\left(1 + b + \frac{b^2}{2} + \ldots\right) = 1 + a + b + \frac{a^2}{2} + \frac{b^2}{2} + ab + \ldots = 1 + (a + b) + \frac{a^2 + 2ab + b^2}{2} + \ldots = 1 + (a + b) + \frac{(a + b)^2}{2} + \ldots = e^{a+b}$

3. $1 - x + \frac{x^2}{2!} - \frac{x^3}{3!} + \ldots$; $e^x \cdot e^{-x} = \left(1 + x + \frac{x^2}{2!} + \frac{x^3}{3!} \ldots\right)\left(1 - x + \frac{x^2}{2!} - \frac{x^3}{3!} + \ldots\right) = 1 + x - x - x^2 + \frac{x^2}{2} + \frac{x^2}{2} + \frac{x^3}{6} - \frac{x^3}{6} + \frac{x^3}{2} - \frac{x^3}{2} \ldots = 1$

4. a. $q!$ has a factor q, which cancels the denominator of $\frac{p}{q}$; the first $q + 1$ terms of the power series all have denominators which are factorials of numbers $\leq q$. Therefore, $q!$ cancels all these, also. **b.** The difference of two integers is an integer.

c. $\frac{1}{q+1} + \frac{1}{(q+1)(q+2)} + \frac{1}{(q+1)(q+2)(q+3)} + \ldots$

d. Since $q + 1 \geq 2$, the first denominator of d is greater than or equal to 2 and every other denominator of d is greater than the corresponding denominator of this sum, so d is strictly smaller than the infinite sum, which equals 1. This shows that d cannot be an integer, which is a contradiction of the assumption that e is rational.

Quiz 1

1. domain: all real numbers; range: $y > 2$

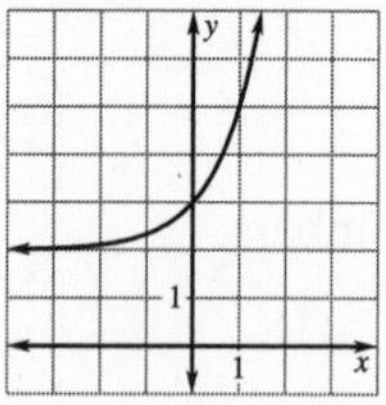

2. domain: all real numbers; range: $y > 2$

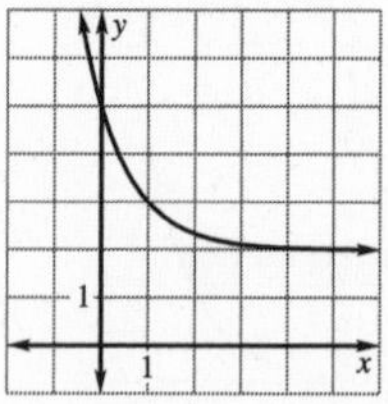

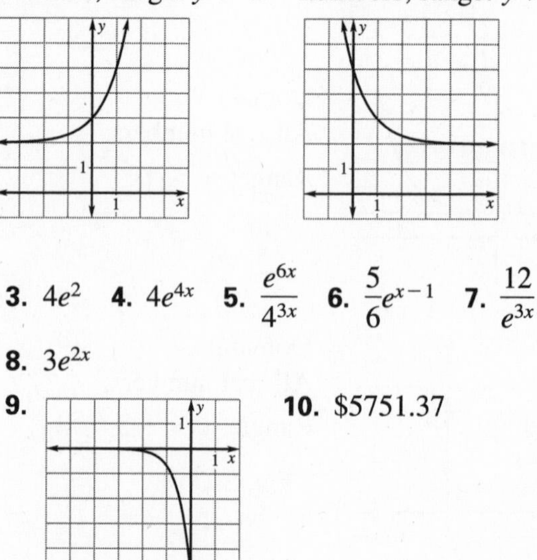

3. $4e^2$ **4.** $4e^{4x}$ **5.** $\frac{e^{6x}}{4^{3x}}$ **6.** $\frac{5}{6}e^{x-1}$ **7.** $\frac{12}{e^{3x}}$

8. $3e^{2x}$

9.

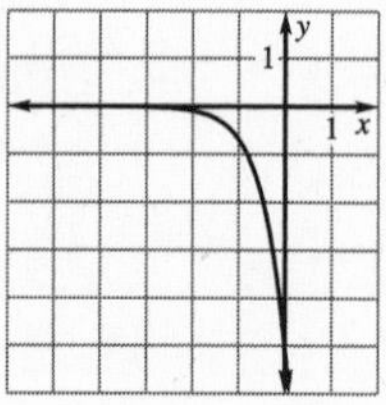

10. $5751.37

Lesson 8.4

Lesson 8.4

Warm-up Exercises

1. 2 **2.** -2 **3.** 1 **4.** -3 **5.** $\frac{2}{3}$

Daily Homework Quiz

1. $2e^{3x}$ **2.** $\dfrac{1}{2e^{4x}}$ **3.** exponential growth

4. exponential decay

5. 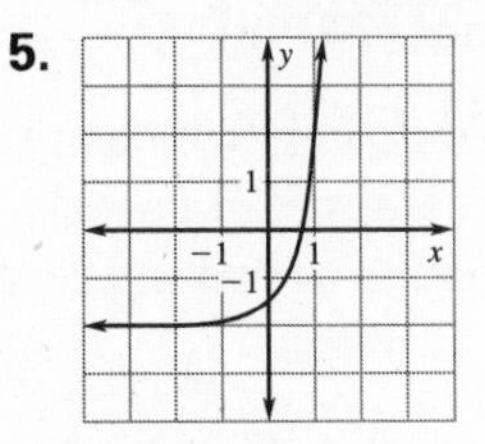

domain: all real numbers; range: $y > -2$

Lesson Opener

Allow 20 minutes.

$\log_2 1 = 0$	$\log_2 2 = 1$	$\log_2 4 = 2$
$\log_2 8 = 3$	$\log_2 16 = 4$	$\log_2 32 = 5$
$\log_3 1 = 0$	$\log_3 3 = 1$	$\log_3 9 = 2$
$\log_3 27 = 3$	$\log_3 81 = 4$	$\log_3 243 = 5$
$\log_4 1 = 0$	$\log_4 4 = 1$	$\log_4 16 = 2$
$\log_4 64 = 3$	$\log_4 256 = 4$	$\log_4 1024 = 5$
$\log_5 1 = 0$	$\log_5 5 = 1$	$\log_5 25 = 2$
$\log_6 6 = 1$	$\log_6 36 = 2$	$\log_6 216 = 3$
$\log_7 7 = 1$	$\log_7 49 = 2$	$\log_{10} 1 = 0$
$\log_{10} 10 = 1$	$\log_{10} 100 = 2$	$\log_{10} 1000 = 3$

Graphing Calculator Activity

1. **3.** **4.** **5.**

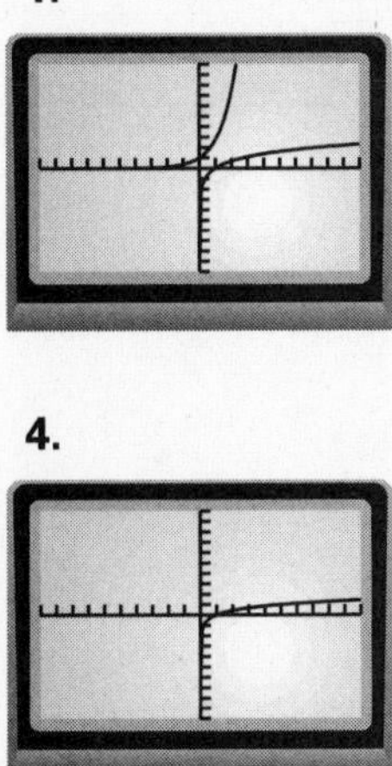

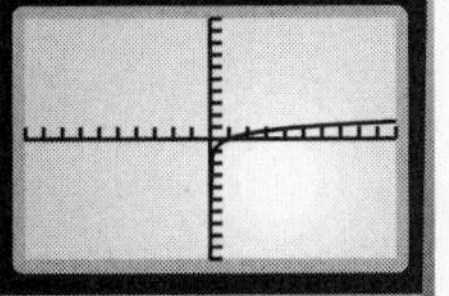

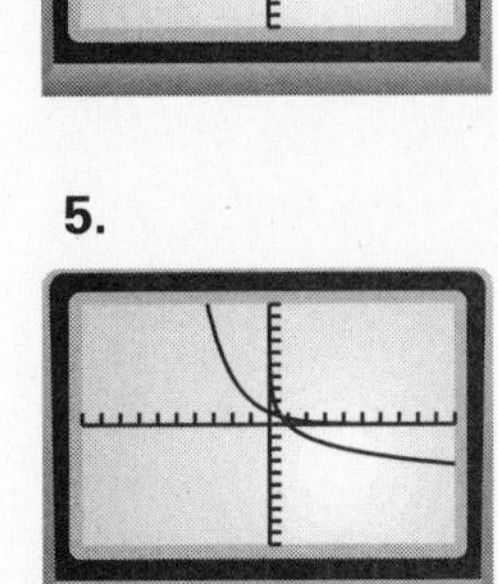

yes

6. **7.** **8.** **9.**

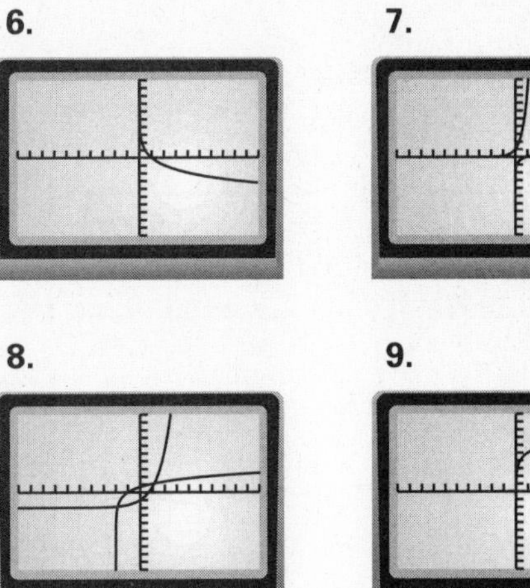

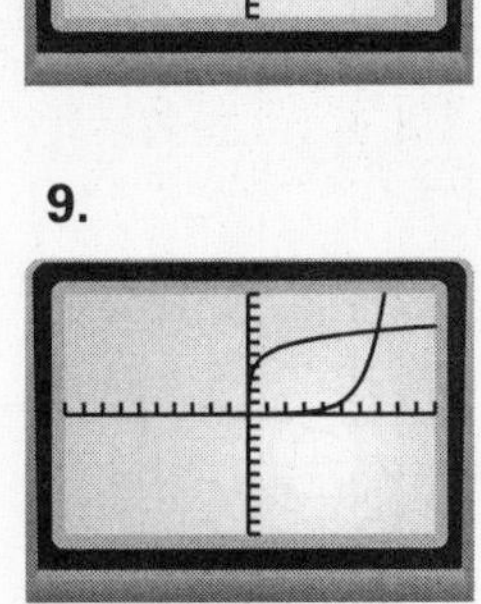

$y = 10^x - 2$ $y = 10^{x-5}$

Practice A

1. $2^3 = 8$ **2.** $5^2 = 25$ **3.** $3^3 = 27$ **4.** $7^2 = 49$ **5.** $2^4 = 16$ **6.** $6^1 = 6$ **7.** 2 **8.** 5 **9.** 2 **10.** 2 **11.** 0 **12.** 1 **13.** 0.778 **14.** -0.398 **15.** 0.571 **16.** 2.079 **17.** -1.470 **18.** 1.812 **19.** x **20.** x **21.** x **22.** x **23.** x **24.** x **25.** A **26.** C **27.** B **28.** C **29.** A **30.** B **31.** 110 decibels

Practice B

1. $4^2 = 16$ **2.** $3^4 = 81$ **3.** $2^0 = 1$ **4.** $9^{1/2} = 3$ **5.** $5^{-1} = \frac{1}{5}$ **6.** $2^{-3} = \frac{1}{8}$ **7.** 0.549 **8.** 1.061 **9.** -0.405 **10.** 3 **11.** 0 **12.** -1 **13.** $\frac{1}{3}$ **14.** $\frac{2}{3}$ **15.** undefined **16.** $f^{-1}(x) = 3^x$ **17.** $f^{-1}(x) = e^x$ **18.** $f^{-1}(x) = \left(\dfrac{1}{3}\right)^x$ **19.** $f^{-1}(x) = \dfrac{10^x}{2}$ **20.** $f^{-1}(x) = 2^x + 1$ **21.** $f^{-1}(x) = 4^{x-2}$

22. 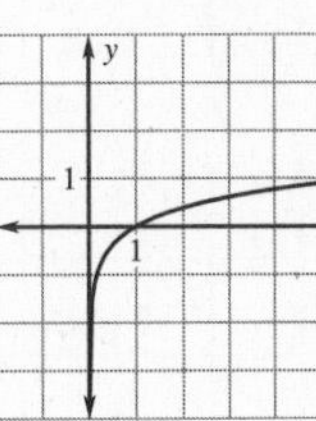

23. 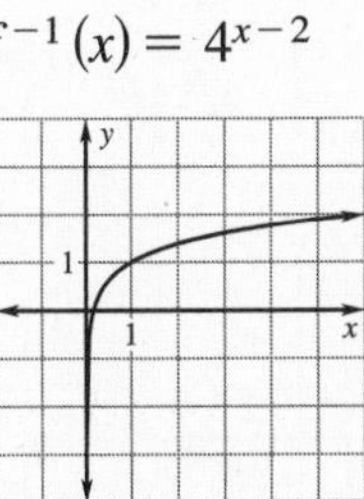

Lesson 8.4 *continued*

24. 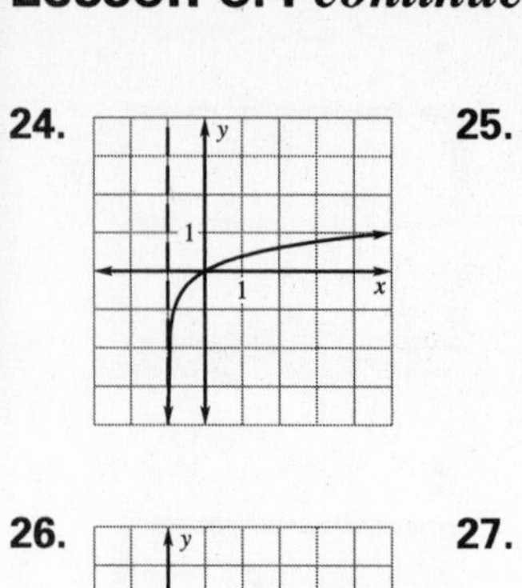

25.

26.

27. 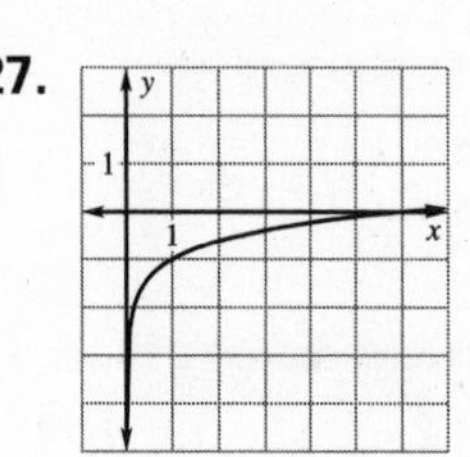

28. $\approx$ 127 strides **29.** $\approx$ 267.4 miles per hour

Practice C

1. $5^3 = 125$ **2.** $8^{1/3} = 2$ **3.** $3^{-3} = \frac{1}{27}$

4. 2.099 **5.** 0.092 **6.** -1.199 **7.** -5

8. -3 **9.** $\frac{2}{3}$ **10.** $\frac{3}{4}$ **11.** $-\frac{2}{3}$ **12.** $-\frac{3}{2}$

13. $f^{-1}(x) = 4^x$ **14.** $f^{-1}(x) = \dfrac{2^x}{7}$

15. $f^{-1}(x) = \dfrac{10^x - 2}{3}$ **16.** $f^{-1}(x) = e^{x+3}$

17. $f^{-1}(x) = e^{x-1} + 2$

18. $f^{-1}(x) = \dfrac{x-4}{2}$ or $f^{-1}(x) = \dfrac{1}{2}x - 2$

19. 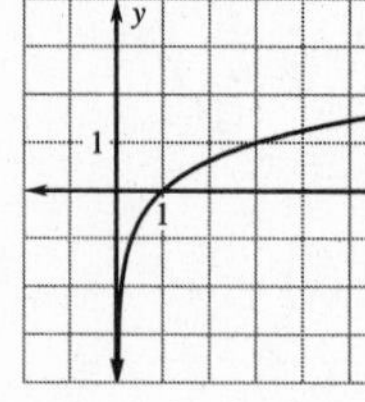

20. 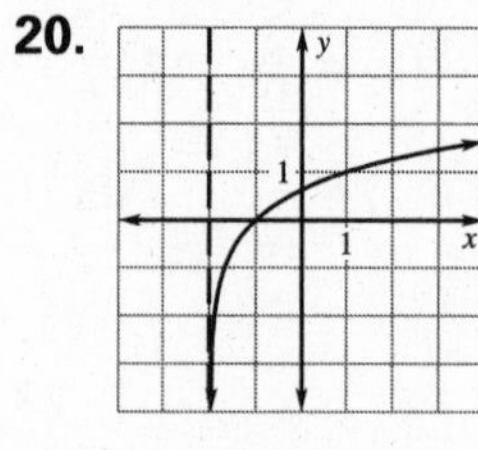

21. 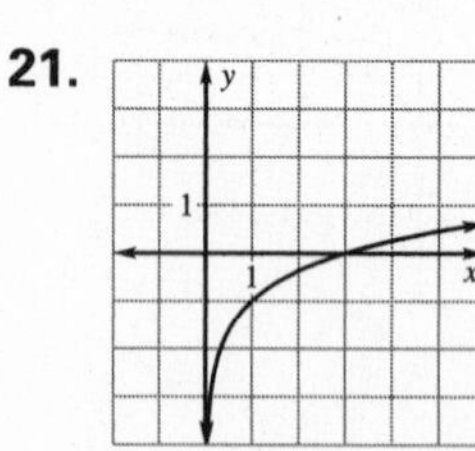

22. 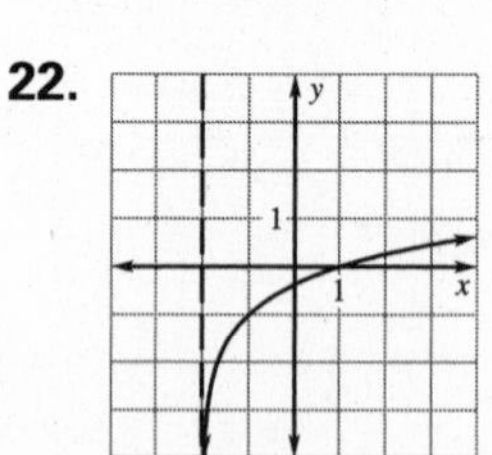

23. 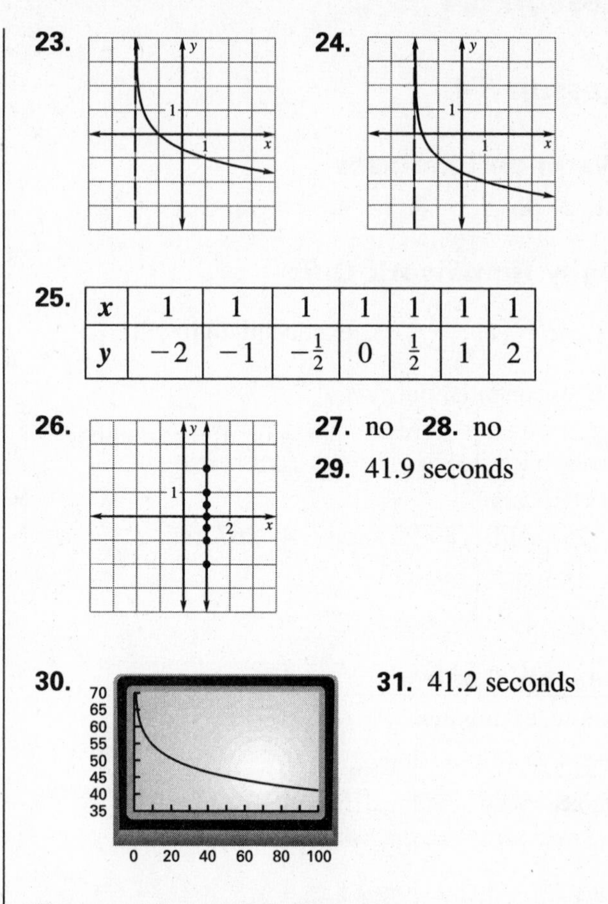

24.

25.

x	1	1	1	1	1	1	1
y	-2	-1	$-\frac{1}{2}$	0	$\frac{1}{2}$	1	2

26.

27. no **28.** no

29. 41.9 seconds

30.

31. 41.2 seconds

Reteaching with Practice

1. $4^4 = 64$ **2.** $5^3 = 125$ **3.** $7^0 = 1$

4. $2^{-3} = \frac{1}{8}$ **5.** $8^1 = 8$ **6.** $\left(\frac{1}{3}\right)^{-1} = 3$ **7.** 5

8. 1 **9.** 0 **10.** $\frac{1}{2}$ **11.** -2 **12.** 5 **13.** x

14. 10 **15.** x **16.** $4x$

17. domain: $x > 0$
range: all real numbers

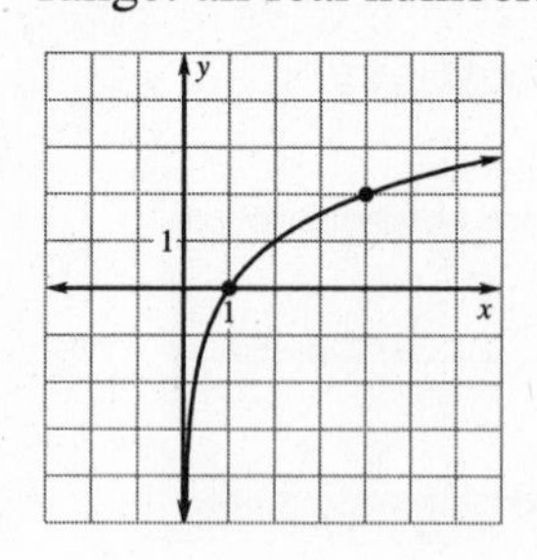

18. domain: $x > 0$
range: all real numbers

Lesson 8.4 *continued*

19. domain: $x > -2$
range: all real numbers

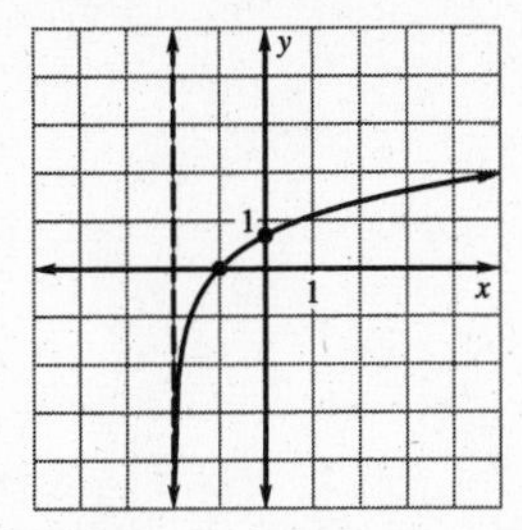

20. domain: $x > 0$
range: all real numbers

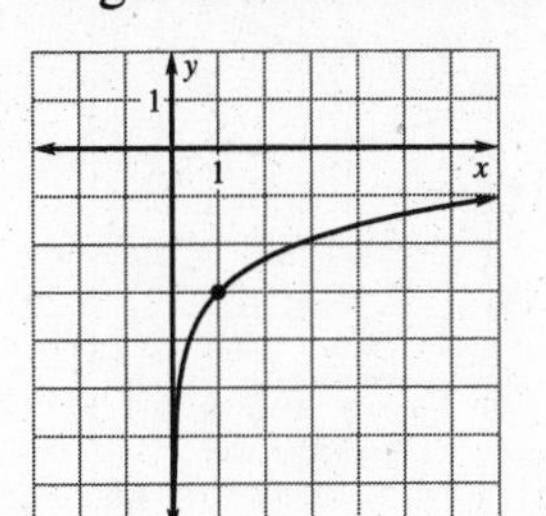

Cooperative Learning Activity

Instructions

2. 283 mi/h **3.** 98 miles **4.** 267 mi/h

5. 15 miles

Analyze the Results

1. The mathematical model was algebraic and problems 2 and 4 were solved in a similar manner, while 3 and 5 were also solved the same way.

2. In problems 2 and 4, you needed to use the Log key; while in 3 and 5, you needed to first use the inverse key, then the Log key.

Interdisciplinary Application

1.

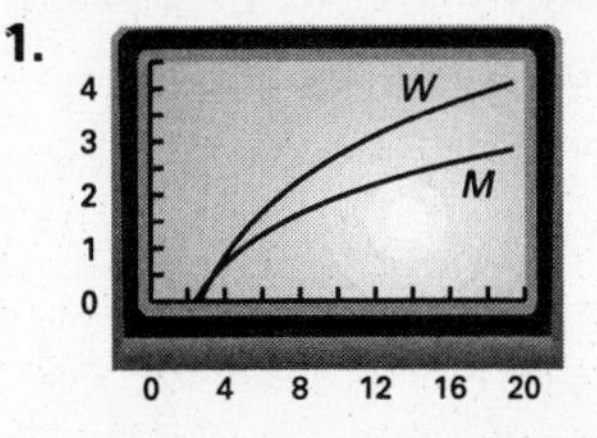

2. Patient A: 0.30; **Patient B:** 0.83; **Patient C:** 1.13; **Patient D:** 0.41; **Patient E:** 0.72; **Patient F:** 1.10

3. Patient C; Patient A

Challenge: Skills and Applications

1. 27 **2.** 8 **3.** $-\frac{5}{3}$ **4.** $\frac{1}{32}$

5. a. $e^x = \left(1 + \frac{1}{k}\right)^{kx} = \left(1 + \frac{1}{\frac{n}{x}}\right)^{\frac{nx}{x}} = \left(1 + \frac{x}{n}\right)^n$

b. $e^{\ln x} = x = \left(1 + \frac{\ln x}{n}\right)^n$; $\ln x = n(\sqrt[n]{x} - 1)$

c. 0.697 **6. a.** 0.8451, 1.8451, 2.8451, 3.8451; 5.8451 **b.** $\log \sqrt{70} = 0.9225$; $\log \sqrt{70} = \frac{1}{2}\log 70$ **c.** 0.4225, 1.4225, 1.9225; $\sqrt{700} = 10\sqrt{7}$; $\sqrt{7000} = 10\sqrt{70}$; In part (a), $70 = 7 \cdot 10$ and $\log 70 = \log 7 + 1$. In part (c) $\sqrt{700} = 10\sqrt{7}$ so $\log \sqrt{700} = \log \sqrt{7} + 1$. Similarly, $\sqrt{7000} = 10\sqrt{70}$, so $\log \sqrt{7000} = \log \sqrt{70} + 1$.

7. $\log_b x = y \Rightarrow b^y = x \Rightarrow \left(\frac{1}{b}\right)^{-y} = x \Rightarrow \log_{1/b} x = -y$

8. ny; $\log_b x = y \Rightarrow b^y = x \Rightarrow (b^{1/n})^{ny} = x \Rightarrow \log_{\sqrt[n]{b}} x = ny$

Lesson 8.5

Warm-up Exercises

1. 6 **2.** 5 **3.** 1 **4.** 4 **5.** 4

Daily Homework Quiz

1. $7^4 = 2401$ **2.** 6 **3.** $4x$ **4.** 2.001

5. $y = e^x + 0.5$

6.

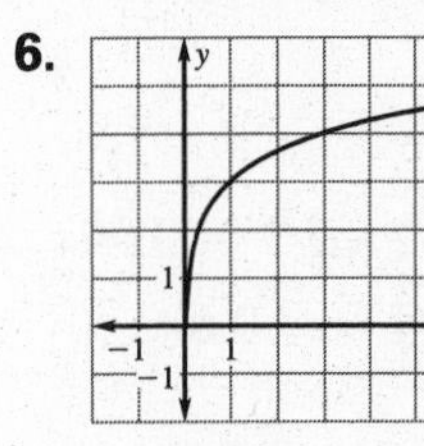

domain: $x > 0$; range: all real numbers

Lesson Opener

Allow 15 minutes.

1. 6 **2.** 8 **3.** 15 **4.** 30 **5.** 20 **6.** 100 **7.** 25 **8.** 70 **9.** ab

Practice A

1. $2 \log 2 \approx 0.602$ **2.** $\log 7 + \log 2 \approx 1.146$

3. $\log 7 - \log 2 \approx 0.544$

4. $\log 2 - \log 7 \approx -0.544$

5. $-3 \log 7 \approx -2.535$ **6.** $2 \log 7 \approx 1.69$

7. $\log_2 3 + \log_2 x$ **8.** $2 + \log_3 x$

9. $\log x - \log 5$ **10.** $1 - \log_6 x$ **11.** $5 \log_3 x$

12. $-3 \ln x$ **13.** $\frac{1}{3} \log x$ **14.** $\frac{1}{2} + \frac{1}{2} \log_2 x$

15. $6 + 2 \log_3 x$ **16.** $\log 15$ **17.** $\log_2 7x$

18. $\log_3 14y$ **19.** $\log\left(\frac{4}{x}\right)$ **20.** $\ln\left(\frac{x}{3}\right)$

21. $\log\left(\frac{x-1}{6}\right)$ **22.** $\ln\left(\frac{2}{x+2}\right)$

23. $\log_3(4x + 20)$ **24.** $\log 8x^2$

Lesson 8.5 *continued*

25. $\dfrac{\log 5}{\log 2} = \dfrac{\ln 5}{\ln 2} \approx 2.322$

26. $\dfrac{\log 10}{\log 7} = \dfrac{\ln 10}{\ln 7} \approx 1.183$

27. $\dfrac{\log 17}{\log 3} = \dfrac{\ln 17}{\ln 3} \approx 2.579$

28. $\dfrac{\log 200}{\log 6} = \dfrac{\ln 200}{\ln 6} \approx 2.957$

29. $\dfrac{\log\left(\frac{1}{2}\right)}{\log 5} = \dfrac{\ln\left(\frac{1}{2}\right)}{\ln 5} \approx -0.431$

30. $\dfrac{\log 1235}{\log 4} = \dfrac{\ln 1235}{\ln 4} \approx 5.135$

31. $t = \dfrac{\ln I - \ln I_0}{0.049}$

32.

I	2000	3000	4000
t	14.1	22.4	28.3

Practice B

1. $\log 3 - \log 4 \approx -0.125$
2. $\log 3 + \log 4 \approx 1.079$ **3.** $2 \log 3 \approx 0.954$
4. $2 \log 4 \approx 1.204$ **5.** $-\log 4 \approx -0.602$
6. $\log 4 - 3 \log 3 \approx -0.829$
7. $\log_6 3 + \log_6 x$ **8.** $\log_2 x - \log_2 5$
9. $\log x + 2 \log y$ **10.** $\log_4 x + \log_4 y - \log_4 3$
11. $\frac{1}{2} \log_3 x + \log_3 y + \log_3 z$
12. $\log_5 2 + \frac{1}{2} \log_5 x$ **13.** $2 \log x - \log 4$
14. $1 - \frac{1}{2} \log x$ **15.** $2 \log_2 x + \log_2 y - \log_2 z$
16. $\log_3\left(\frac{7}{x}\right)$ **17.** $\log_5 3x^2$ **18.** $\log_4 5xy$
19. $\log \dfrac{\sqrt{x}}{4}$ **20.** $\log_2 \dfrac{\sqrt[3]{x^2}}{y^3}$ **21.** $\log_3\left(\dfrac{4x^2}{5}\right)$
22. $\dfrac{\log 12}{\log 3} = \dfrac{\ln 12}{\ln 3} \approx 2.262$
23. $\dfrac{\log 2}{\log 6} = \dfrac{\ln 2}{\ln 6} \approx 0.387$
24. $\dfrac{\log 0.5}{\log 4} = \dfrac{\ln 0.5}{\ln 4} = -0.5$
25. $\dfrac{\log 12}{\log 0.8} = \dfrac{\ln 12}{\ln 0.8} \approx -11.136$
26. $\dfrac{\log 2.8}{\log 1.5} = \dfrac{\ln 2.8}{\ln 1.5} \approx 2.539$
27. $\dfrac{\log 6}{\log \frac{1}{2}} = \dfrac{\ln 6}{\ln \frac{1}{2}} \approx -2.585$
28. $\text{pH} = 6.1 + \log B - \log C$ **29.** ≈ 7.2
30. below normal
31. $\text{pH} = 7.48 - \log C$ **32.** 1.2

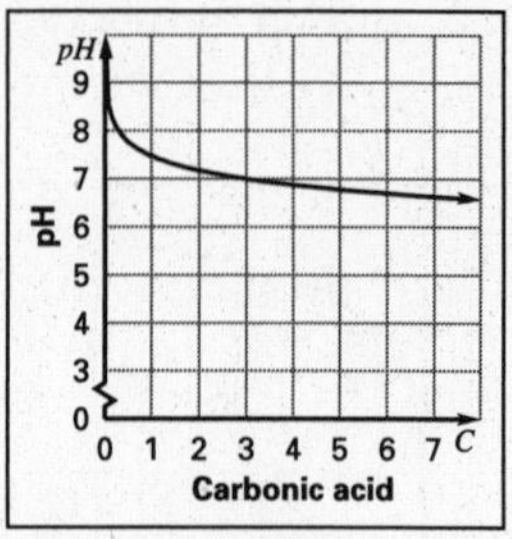

Practice C

1. $\ln 2 + \ln 3 \approx 1.792$
2. $\ln 2 + \ln 5 - \ln 3 \approx 1.203$
3. $\ln 2 + \ln 3 + \ln 5 \approx 3.401$
4. $2 \ln 2 + \ln 3 \approx 2.485$
5. $\ln 2 - \ln 5 \approx -0.916$
6. $\ln 5 - \ln 3 - \ln 2 \approx -0.183$
7. $\log 8 + \log x$ **8.** $\log_3 x + \log_3 y + \log_3 z$
9. $\frac{1}{2} + \log_4 x + \log_4 y - \log_4 z$
10. $\ln x - \ln y - \ln z$
11. $\frac{1}{2}(\log 3 + \log x + \log y)$
12. $\frac{1}{2} \log_5 x - \log_5 y$ **13.** $\ln 3 + \ln y - \frac{1}{4} \ln x$
14. $3(\log 3 + \log x + \log y + 2 \log z)$
15. $4(\log_2 x + \log_2 y) - 2 \log_2 z$
16. $\log \dfrac{3}{28}$ **17.** $\ln \dfrac{3xz}{y}$ **18.** $\ln \dfrac{x^3}{y^2 z^4}$
19. $\log_2 \dfrac{(x-4)(x+1)^5}{(x-1)^3}$
20. $\log_2 \dfrac{\sqrt{x+5}}{x^2} + \ln y$
21. $\ln\left[\dfrac{(x-2)(x+1)^2}{(x+2)(x-1)^5}\right]^3$
22. $y = \dfrac{\log x}{\log 3}$ or $y = \dfrac{\ln x}{\ln 3}$
23. $y = \dfrac{\log(x+3)}{\log 6}$ or $y = \dfrac{\ln(x+3)}{\ln 6}$
24. $y = \dfrac{\log(x-1)}{\log 2} + 3$ or $y = \dfrac{\ln(x-1)}{\ln 2} + 3$
25. $t = \dfrac{\ln(Sr + Pn) - \ln P - \ln n}{n(\ln(n+r) - \ln n)}$
26. ≈ 19.7 years

Lesson 8.5 *continued*

Reteaching with Practice

1. 2.059 **2.** 2.57 **3.** 0.511 **4.** 2.322
5. $\log 9 + \log x$ **6.** $\log_2 6 + 3 \log_2 x$
7. $\log_6 2 - \log_6 3$ **8.** $\log_3 4 + \log_3 x - \log_3 5$
9. $\ln 2 + \ln x + \ln y$ **10.** $\ln 2 + 2 \ln x - \ln y$
11. $\log_4 60$ **12.** $\log \frac{x}{y}$ **13.** $\ln 2$
14. $\log_2 27$ **15.** $\log_2 x^9$ **16.** $\ln 3$ **17.** 3.096
18. 1.850 **19.** 4.087 **20.** 1.431

Real-Life Application

1. 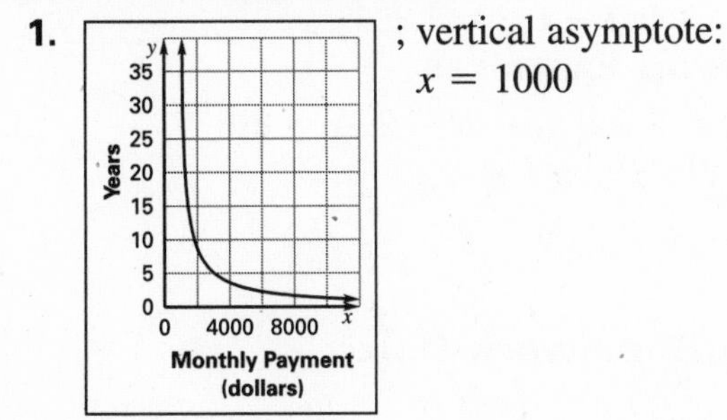

; vertical asymptote: $x = 1000$

2. $12.542(\ln x - \ln(x - 1000))$
3. about 25 years **4.** about 30 years
5. \$347,316; \$197,316

Math and History

1. The first question is: "Is your number less than 500,000?" It should take 20 guesses.
2. The choice of one out of 94 key codes contains about 6.6 bits of information, so it will take 7 bits to code this part of the keyboard, which is less than one byte. The entire 104-key extended keyboard, shifted and unshifted, can be coded in one byte. **3.** $\log_2 300{,}000{,}000$ is about 28.2, so 29 bits will do it.

Challenge: Skills and Applications

1. $2q$ **2.** $-(p + q)$ **3.** $\frac{p}{2}$ **4.** $2p + q$
5. $q - 1$ **6.** $p + 2$
7. a. 0.602, 0.699, 0.903 **b.** 0.476, 0.777, 0.952 **c.** 0.845 **8. a.** $b^x = u$; $b^y = c$
b. $b = c^{1/y}$; $(c^{1/y})^x = u \Rightarrow c^{x/y} = u \Rightarrow \log_c u = \frac{x}{y} = \frac{\log_b u}{\log_b c}$ **9.** $\log_b a = \frac{\log_a a}{\log_a b} = \frac{1}{\log_a b}$

10. a. -139, 417, 834
b. $\log_g (64 \cdot 128) = \log_g 64 + \log_g 128 = 834 + 973 = 1807$; therefore, $64 \cdot 128 = (1.005)^{1807} \approx 8205$ (actual value: 8192)

Lesson 8.6

Warm-up Exercises

1. 3^{12x} **2.** 0.861 **3.** 4^{4x-6} **4.** $\log_3 5$ **5.** 3

Daily Homework Quiz

1. 12 **2.** -2.176 **3.** $\frac{1}{4}\ln 7 + \ln 2 + \frac{2}{3}\ln x$
4. $\log_5 4$ **5.** 1.441

Lesson Opener

Allow 15 minutes.
1. 1.05 **2.** 0, 1.85 **3.** 2.76 **4.** -2.64, 4.64
5. no solutions **6.** 2.73, 5.05 **7.** 133.14
8. 0.01, 8.73

Practice A

1. yes **2.** no **3.** no **4.** no **5.** yes **6.** no
7. no **8.** yes **9.** no **10.** yes **11.** no
12. yes **13.** -1 **14.** -5 **15.** -2 **16.** 7
17. $\frac{4}{3}$ **18.** $\frac{7}{4}$ **19.** $\log_2 9$ **20.** $\log_3 10$ **21.** $\ln 5$
22. $\frac{\ln 6}{2}$ **23.** $\log_2 7$ **24.** $\frac{\log_5 10}{3}$ **25.** 7 **26.** 7
27. 3 **28.** -6 **29.** $\frac{3}{4}$ **30.** $-\frac{3}{2}$ **31.** 32
32. 6562 **33.** 499,998.5 **34.** $\frac{e^2 + 3}{5}$ **35.** 0
36. 25 **37.** 3.2 years **38.** 13.5 years
39. 23.1 years

Practice B

1. 2.890 **2.** 2.544 **3.** 1.869 **4.** 1.609
5. 1.585 **6.** 0.646 **7.** 0.667 **8.** 0.805
9. 0.886 **10.** 0.462 **11.** 0.576
12. -2.322 **13.** -0.5 **14.** -0.973
15. -1.946 **16.** 1.609 **17.** 2 **18.** -1.792
19. 0.229 **20.** 0.308 **21.** 0 **22.** 0.347
23. -1.099 **24.** 25.850 **25.** 2.485
26. 1.445 **27.** 1.528 **28.** 148.413 **29.** 0.01
30. 2.828 **31.** 20.086 **32.** 100,000 **33.** 0.001

Lesson 8.6 *continued*

34. 2980.958 **35.** 20.086 **36.** 148.413
37. 10,000 **38.** 46.416 **39.** 3 **40.** 0.4
41. 0.002 **42.** 300,651.071 **43.** 21.333 **44.** 1
45. no solution **46.** 1.5 **47.** no solution
48. 3 **49.** 11.185 years **50.** 20.086

Practice C

1. 2.197 **2.** 0.333 **3.** 3.386 **4.** 0.349
5. 1.436 **6.** 6.447 **7.** 0.376 **8.** 1.269
9. 0.258 **10.** 0 **11.** $-1, 2$ **12.** $-1, 0.667$
13. 4.5 **14.** 22,023.466 **15.** 11 **16.** 181.939
17. 2.414 **18.** 1 **19.** 3 **20.** 0.143 **21.** 3.333
22. no solution **23.** $-2, 3$ **24.** 7 **25.** 4, 6
26. no solution **27.** 0.461 **28.** 3.697
29. -8.266 **30.** no solution **31.** 5.303
32. 7.193 **33.** 5.2 years **34.** 30 years
35. $211,320 **36.** $131,320

Reteaching with Practice

1. 4 **2.** 2 **3.** -1 **4.** 1 **5.** 1.292
6. -1.609 **7.** 2 **8.** 1.091 **9.** 0
10. -1.242 **11.** 1 **12.** -2 **13.** 3 **14.** 9
15. 23 **16.** 1.745

Interdisciplinary Application

1. $\text{pH} = 6.1 + \log_{10}x - \log_{10}y$ **2.** 7.426; yes
3. 27.99 **4.** 1.30

Challenge: Skills and Applications

1. a. $y = \log_{x^2} a \Rightarrow (x^2)^y = a \Rightarrow x^{2y} = a$
$\Rightarrow \log_x a = 2y \Rightarrow 2\log_{x^2} a = \log_x a$
$\Rightarrow \log_{x^2} a = \frac{1}{2}\log_x a$

b. $\log_{x^n} a = \dfrac{1}{n}\log_x a$ **2.** $3, -1$ **3.** $2, \frac{1}{2}$ **4.** 6

5. 8 **6.** $\sqrt[5]{6}$ **7.** $\frac{1}{12}$ **8. a.** $2P = P(1 + r)^T$

b. $T = \dfrac{\ln 2}{\ln(1 + r)} \approx \dfrac{\ln 2}{r}$ **c.** $\ln 2 \approx 0.70$

9. a. $A = A_0\left(\frac{1}{2}\right)^{t/h}$ **b.** 14.7 minutes

Quiz 2

1. 4 **2.** -2 **3.** $\frac{1}{2}$

4.

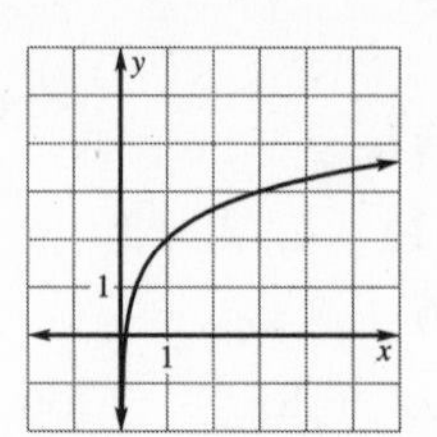

5.

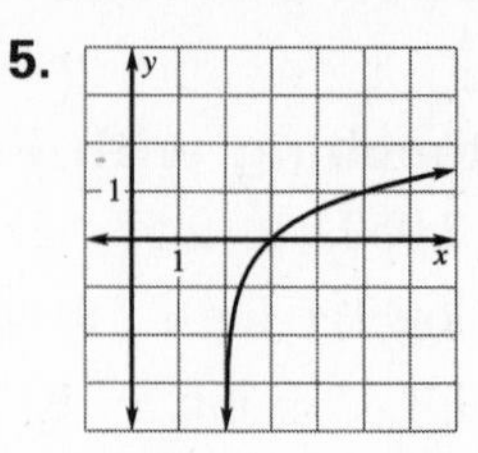

6. 3 **7.** -1 **8.** $\frac{1}{3}\log_5 x + 6\log_5 y$
9. $\log_3 135x^2$ **10.** 1.948 **11.** 1.386
12. 27.86 **13.** no solution **14.** 3
15. 13.86 years

Lesson 8.7

Warm-up Exercises

1. $x_1 = 3$ and $y_2 = 50$ **2.** $y = ab^x$
3. $\log_4 2.7 = b$ **4.** $64 \cdot 2^x$
5. about -2.569

Daily Homework Quiz

1. yes **2.** no **3.** -6 **4.** $\dfrac{\ln 3}{3} \approx 0.366$

5. no solution **6.** $2e^{-4/3} \approx 0.527$

Lesson Opener

Allow 15 minutes.

1.

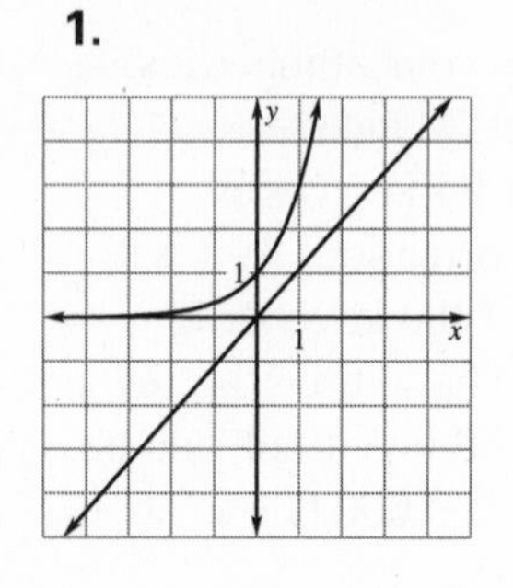

2.

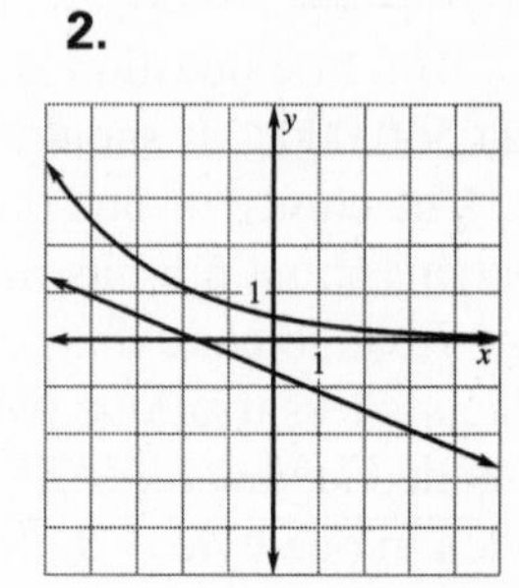

3.

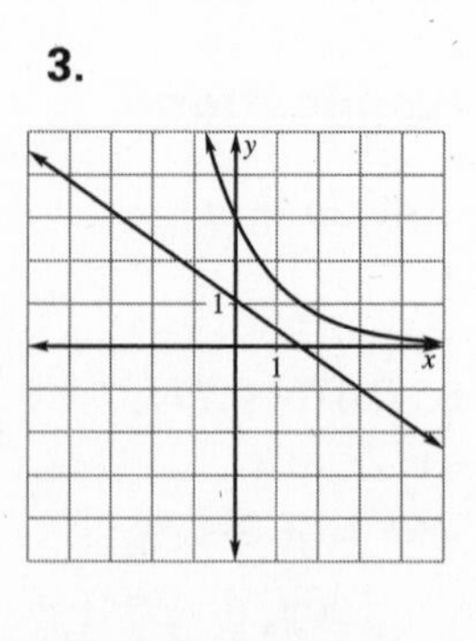

4.

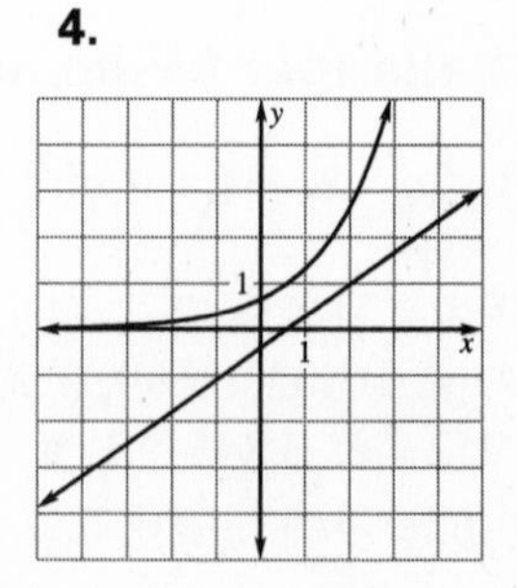

5. linear function

Practice A

1. $y = 3^x$ **2.** $y = 3 \cdot 2^x$ **3.** $y = 2 \cdot 5^x$
4. $y = \left(\frac{1}{2}\right)4^x$ **5.** $y = 2^x$ **6.** $y = 2 \cdot 3^x$
7. yes **8.** no **9.** yes **10.** yes
11. $y = 4.96(1.38)^t$ **12.** $y = 1.52(3.33)^t$
13. $y = 171.40(186{,}278.85)^t$
14. $y = 3.10(24.70)^t$ **15.** $y = 3459.92(2.81)^t$
16. $y = 5.07(9.98)^t$ **17.** $y = 2x^3$ **18.** $y = 3x^2$
19. $y = x^{1.5}$ **20.** no **21.** yes **22.** $y = x^{2.4}$
23. $y = x^{1.3}$ **24.** $y = x^{0.8}$

Practice B

1. $y = \left(\frac{1}{3}\right)2^x$ **2.** $y = \left(\frac{1}{25}\right)4^x$ **3.** $y = 3\left(\frac{1}{2}\right)^x$
4. $y = 6\left(\frac{2}{3}\right)^x$ **5.** $y = \left(\frac{1}{2}\right)5^x$ **6.** $y = \frac{5}{4}\left(\frac{1}{3}\right)^x$

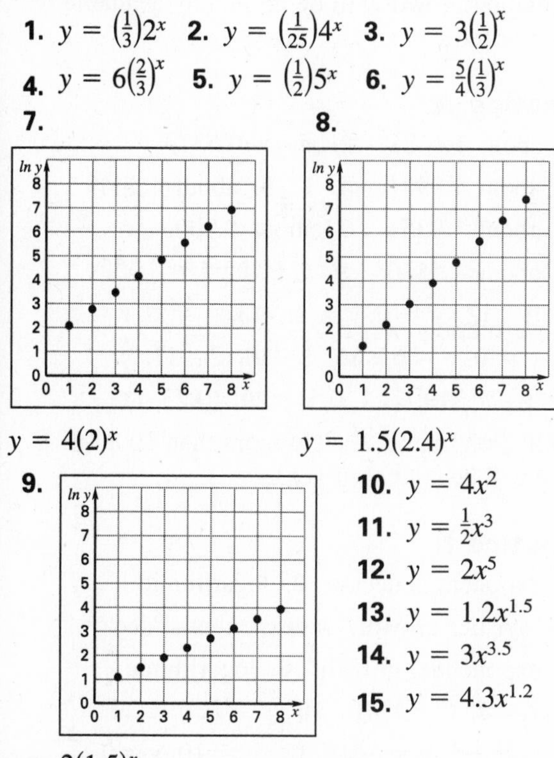

7. $y = 4(2)^x$ **8.** $y = 1.5(2.4)^x$

9. $y = 2(1.5)^x$

10. $y = 4x^2$
11. $y = \frac{1}{2}x^3$
12. $y = 2x^5$
13. $y = 1.2x^{1.5}$
14. $y = 3x^{3.5}$
15. $y = 4.3x^{1.2}$

16. **17.**

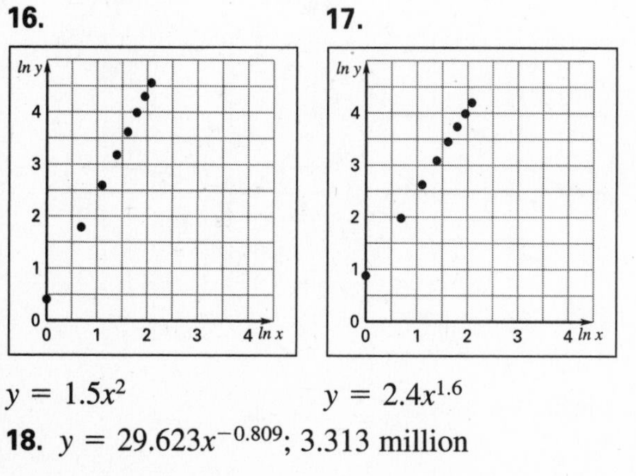

$y = 1.5x^2$ $y = 2.4x^{1.6}$

18. $y = 29.623x^{-0.809}$; 3.313 million

Practice C

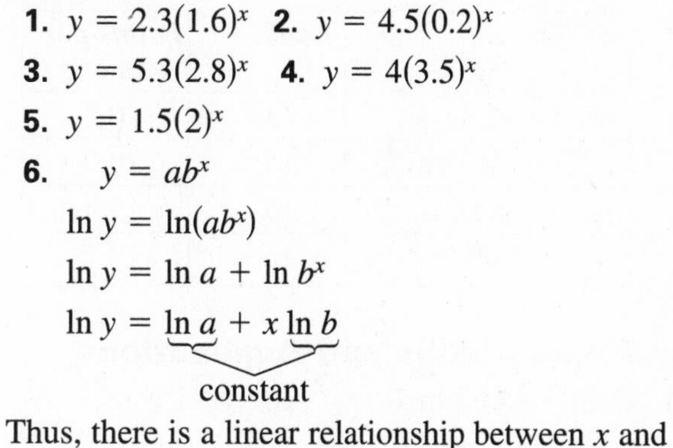

1. $y = 2.3(1.6)^x$ **2.** $y = 4.5(0.2)^x$
3. $y = 5.3(2.8)^x$ **4.** $y = 4(3.5)^x$
5. $y = 1.5(2)^x$
6.
$$\begin{aligned} y &= ab^x \\ \ln y &= \ln(ab^x) \\ \ln y &= \ln a + \ln b^x \\ \ln y &= \underbrace{\ln a}_{\text{constant}} + x\,\underbrace{\ln b}_{\text{constant}} \end{aligned}$$

Thus, there is a linear relationship between x and $\ln y$. **7.** $y = 1.5x^{0.5}$ **8.** $y = 2.4x^{1.5}$
9. $y = 8.3x^{0.25}$ **10.** $y = 3x^{1.2}$ **11.** $y = 2.5x^{2.5}$
12.
$$\begin{aligned} y &= ax^b \\ \ln y &= \ln ax^b \\ \ln y &= \ln a + \ln x^b \\ \ln y &= \underbrace{\ln a}_{\text{constant}} + b\,\ln x \end{aligned}$$

Thus, there is a linear relationship between $\ln x$ and $\ln y$. **13.** $y = 28.38(1.14)^x$
14. $y = 30.84x^{0.33}$ **15.** The exponential model is better because the relationship between x and $\ln y$ is closer to linear than the relationship between $\ln x$ and $\ln y$.

Reteaching with Practice

1. $y = 7 \cdot 2^x$ **2.** $y = -6 \cdot 2^x$ **3.** $y = 3 \cdot 3^x$
4. $y = 0.25x^2$ **5.** $y = 0.75x^3$ **6.** $y = 2x^{0.5}$

Real-Life Application

1.

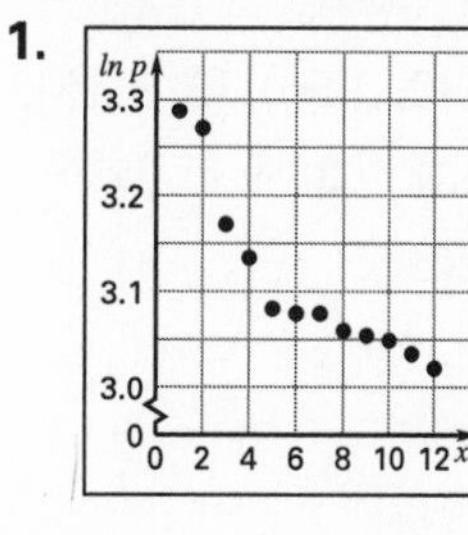

2.

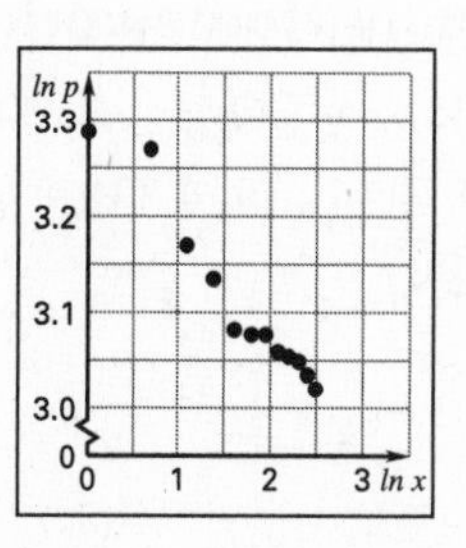

3. Power model; points follow more of a linear pattern. **4.** $p = 27.15x^{-0.115}$

Lesson 8.7 *continued*

5.

Player	*Rank, x*	*Points per game, p*
Michael Finley	13	20.2
Alonzo Mourning	14	20.0
Kobe Bryant	15	19.9
Mitch Richmond	16	19.7

Challenge: Skills and Applications

1. e^3, e^{-2} **2.** $-\ln 2$

3. a. $k = \log_c b = \dfrac{\log b}{\log c} = \dfrac{\ln b}{\ln c}$

b. $y = 100e^{x \ln(2/3)}$; $y = 100\left(\frac{1}{2}\right)^{x \log(2/3)/\log(1/2)}$, or $y = 100\left(\frac{1}{2}\right)^{x \ln(2/3)/\ln(1/2)}$

4. a. $m = f(x_0) \leftrightarrow (\ln b)ab^{x_0} = ab^{x_0} \leftrightarrow \ln b = 1 \leftrightarrow b = e$ **b.** $y = \dfrac{4}{\ln 3} \cdot 3^x$

5. a. $y = ab^{x^2}$ **b.** $y = \dfrac{1}{\sqrt{2\pi}}e^{-x^2/2}$

c. The function approaches 0 as $x \to \infty$, and also as $x \to -\infty$. The graph has a turning point at $x = 0$.

Lesson 8.8

Warm-up Exercises

1. about 3.05 **2.** about 0.23 **3.** about 1 **4.** $y = 1$ **5.** about 0.402

Daily Homework Quiz

1. $y = 36\left(\frac{2}{3}\right)^x$ **2.** (1, 1.97), (2, 2.56), (3, 3.15), (4, 3.74), (5, 4.33); $y = 1.7(1.2)^x$ **3.** $y = 2x^{3/2}$

4.

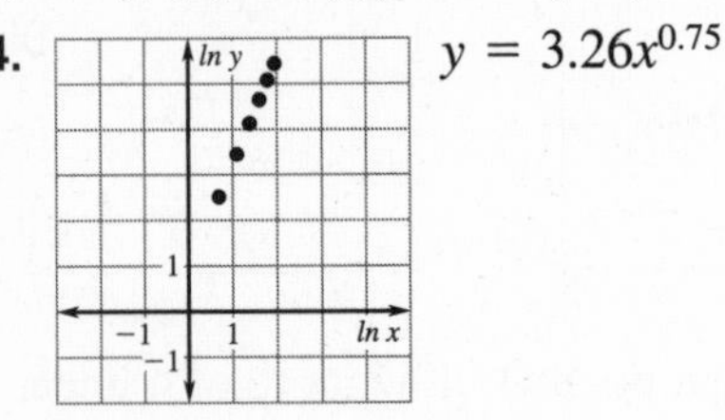

$y = 3.26x^{0.75}$

Lesson Opener

Allow 15 minutes.

Note: The answers to problems 3 and 6 were calculated using unrounded constants in the regression equations. Student answers may vary somewhat due to rounding.

1. $P = 26.307(1.145)^t$

2, 5.

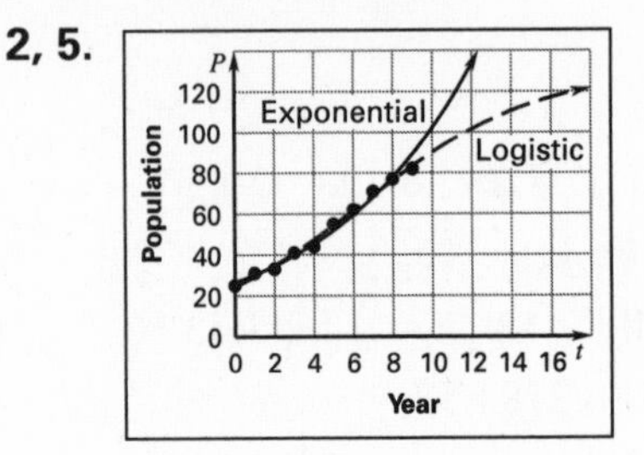

3. 395; 1532; 23,012 **4.** $P = \dfrac{130.673}{1 + 4.372e^{-0.227t}}$

6. 125; 130; 131 **7.** *Sample answer:* The logistic model is better. The exponential model is unreasonable because eventually the rabbit population growth will be limited by available space or food.

Practice A

1. about 1.4621 **2.** about 0.5379 **3.** about 1.9951 **4.** 1 **5.** about 1.2449 **6.** about 1.9354 **7.** about 0.9003 **8.** about 1.5546 **9.** C **10.** A **11.** B **12.** $y = 0, y = 1$ **13.** $y = 0, y = 5$ **14.** $y = 0, y = 6$ **15.** $\frac{1}{3}$ **16.** 2 **17.** $\frac{5}{2}$ **18.** (0, 2) **19.** (1.1, 0.5) **20.** (0.23, 1) **21.** 89,963 units **22.** No more than 100,000 units will be sold each year.

Practice B

1. exponential decay **2.** logarithmic **3.** logistics growth **4.** exponential decay **5.** exponential growth **6.** logarithmic **7.** A **8.** C **9.** B **10.** $y = 0, y = 20$ **11.** $y = -5, y = -4$ **12.** $y = 10, y = 12$

13.

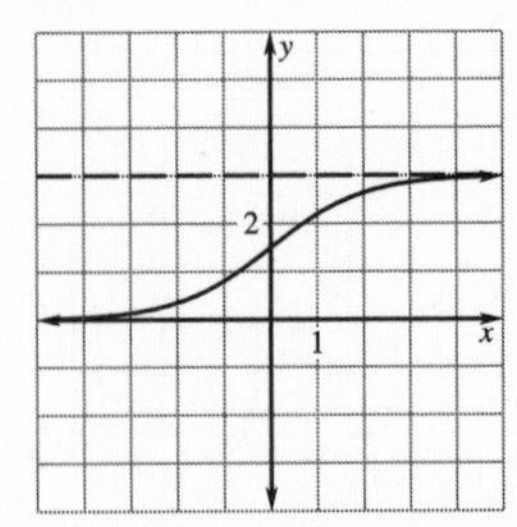

14.

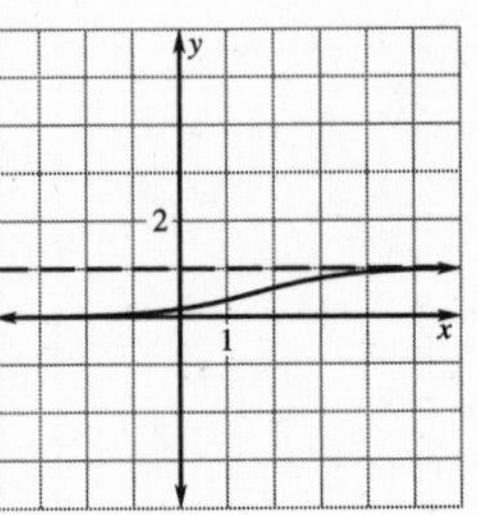

Lesson 8.8 *continued*

15. 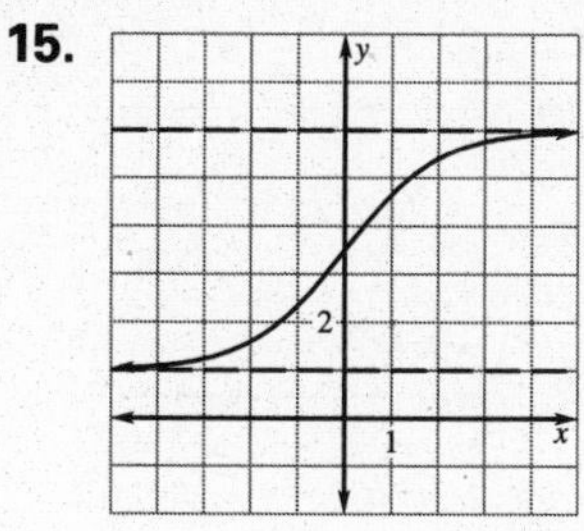

16. $\ln 2 \approx 0.693$

17. $\ln \frac{5}{3} \approx 0.511$ 18. $\frac{1}{2} \ln 5 \approx 0.805$

19.

20. $y = 0, y = 500$

21. 500

22. ≈ 451

Practice C

1. 2.667 2. 7.276 3. 0.194 4. 8
5. 0.0000012 6. 4.993 7. 0.400 8. 7.985

9.–14.

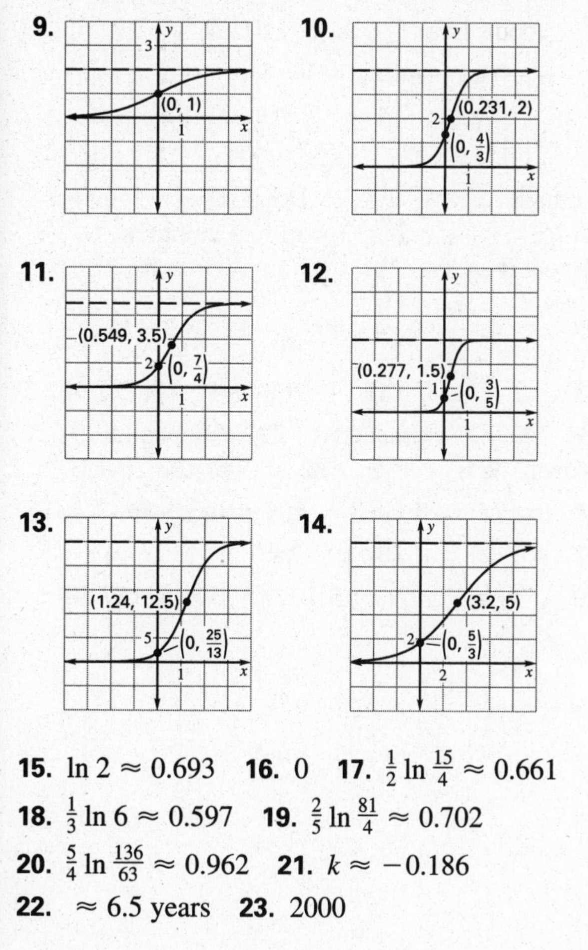

15. $\ln 2 \approx 0.693$ 16. 0 17. $\frac{1}{2} \ln \frac{15}{4} \approx 0.661$
18. $\frac{1}{3} \ln 6 \approx 0.597$ 19. $\frac{2}{5} \ln \frac{81}{4} \approx 0.702$
20. $\frac{5}{4} \ln \frac{136}{63} \approx 0.962$ 21. $k \approx -0.186$
22. ≈ 6.5 years 23. 2000

24. $y = \dfrac{c}{1 + ae^{-r(0)}}$

$y = \dfrac{c}{1 + ae^{0}}$

$y = \dfrac{c}{1 + a}$

25. $(-rx) \to -\infty$

26. $e^{-rx} \to 0$ 27. $ae^{-rx} \to 0$ 28. $1 + ae^{-rx} \to 1$

29. $\dfrac{c}{1 + ae^{-rx}} \to c$

Reteaching with Practice

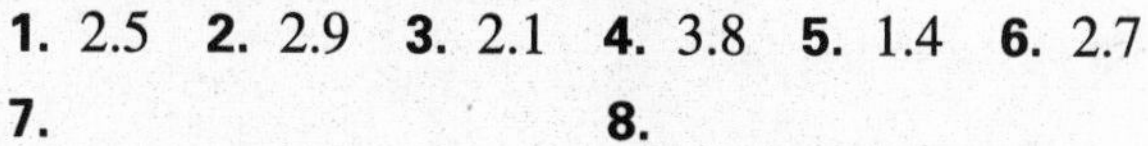

1. 2.5 2. 2.9 3. 2.1 4. 3.8 5. 1.4 6. 2.7

7. 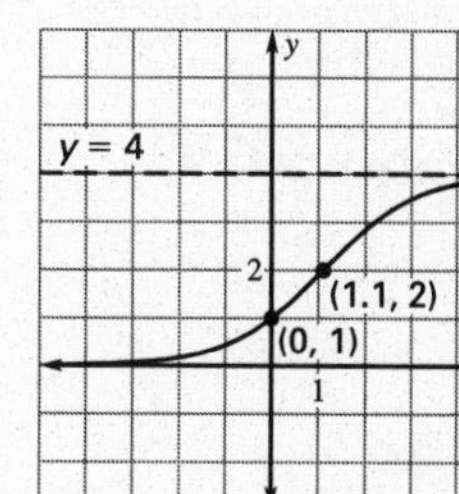

8. 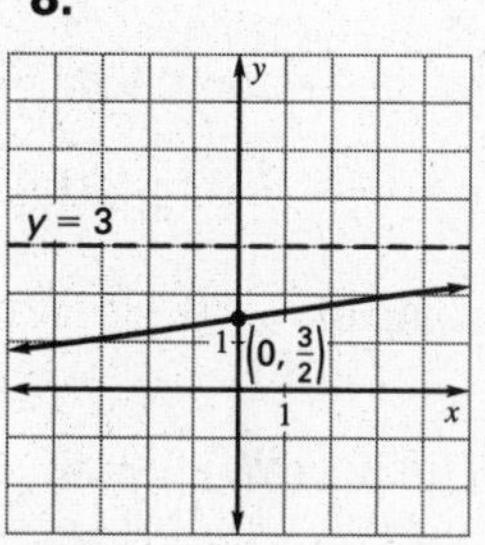

9. 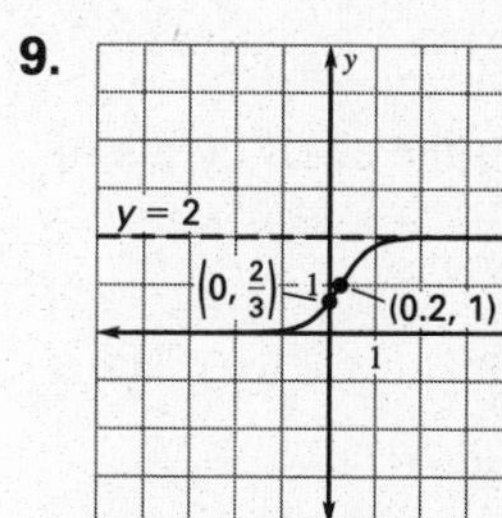

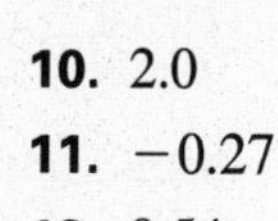

10. 2.0
11. −0.27

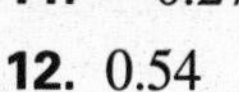

12. 0.54

Interdisciplinary Application

1. 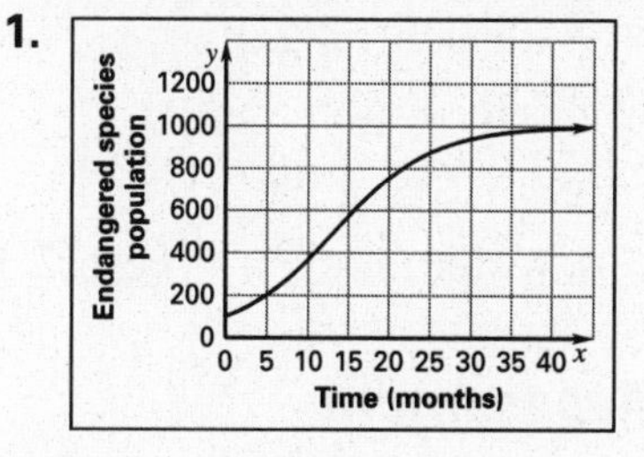

horizontal asymptotes: $y = 0, y = 1000$

2. about 295; about 530 3. 17 months

4. $P = \dfrac{719.36}{1 + 8.34e^{-0.2885t}}$

Challenge: Skills and Applications

1. $y = \dfrac{8}{1 + 9e^{-(\ln 3)x}}$

Lesson 8.8 *continued*

2.

a. $f(3) = \dfrac{8}{1 + 9e^{-3(\ln 3)}} = \dfrac{8}{1 + 9\left(\frac{1}{27}\right)} = \dfrac{8}{\frac{4}{3}} = 6$

b. The graph has rotational symmetry about the point of maximum growth (2, 4).

3. $f(x) - \dfrac{c}{2} = \dfrac{c}{1 + e^{-rx}} - \dfrac{c}{2} = \dfrac{c - ce^{-rx}}{2(1 + e^{-rx})} =$

$\dfrac{ce^{rx} - c}{2(e^{rx} + 1)}$ (after multiplying top and bottom by e^{rx}); $\dfrac{c}{2} - f(-x) = \dfrac{c}{2} - \dfrac{c}{1 + e^{rx}} = \dfrac{ce^{rx} - c}{2(e^{rx} + 1)}$

4. a. As y approaches c, R approaches $\dfrac{r}{c} \cdot c \cdot 0 = 0$; similarly, as y approaches 0, R approaches $\dfrac{r}{c} \cdot 0 \cdot c = 0$.

b. The maximum point of the graph of $R = \dfrac{r}{c}x(c - x) = -\dfrac{r}{c}x^2 + rx$ is at $x = -\dfrac{b}{2a} =$

$\dfrac{r}{2(r/c)} = \dfrac{c}{2}$ where a and b are the coefficients of the parabola in standard form.

5. a. $[\cosh(x)]^2 - [\sinh(x)]^2 = \frac{1}{4}(e^x + e^{-x})^2 - \frac{1}{4}(e^x - e^{-x})^2 = \frac{1}{4}(e^{2x} + 2 + e^{-2x}) - \frac{1}{4}(e^{2x} - 2 + e^{-2x}) = 1$

b. (i) $2\cosh(x)\sinh(x) = 2 \cdot \frac{1}{2}(e^x + e^{-x}) \cdot \frac{1}{2}(e^x - e^{-x}) = \frac{1}{2}(e^{2x} - e^{-2x}) = \sin(2x)$

(ii) $[\cosh(x)]^2 + [\sinh(x)]^2 = \frac{1}{4}(e^x + e^{-x})^2 + \frac{1}{4}(e^x - e^{-x})^2 = \frac{1}{4}(e^{2x} + 2 + e^{-2x}) + \frac{1}{4}(e^{2x} - 2 + e^{-2x}) = \frac{1}{2}(e^{2x} + e^{-2x}) = \cosh(2x)$

Review and Assessment

Review Games and Activities

1. a; A **2.** b, c; B, C **3.** e; E **4.** d; D

Test A

Graph 1–6 on graph paper

1.

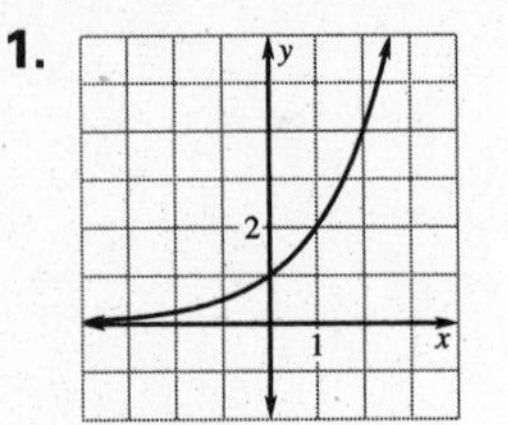

Domain: all real numbers
Range: $y > 0$

2.

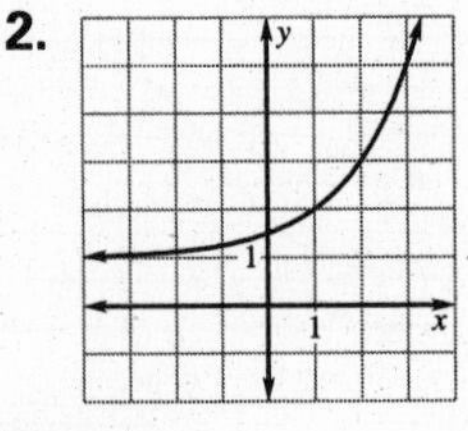

Domain: all real numbers
Range: $y > 1$

3.

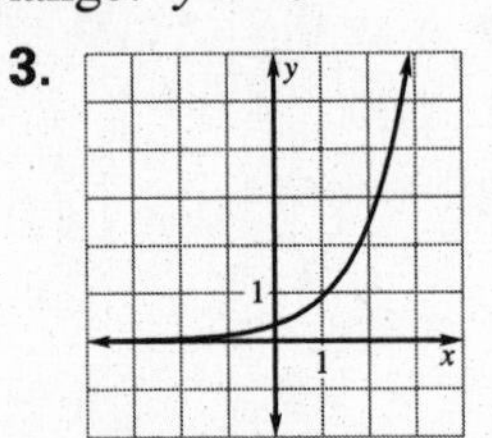

Domain: all real numbers
Range: $y > 0$

4.

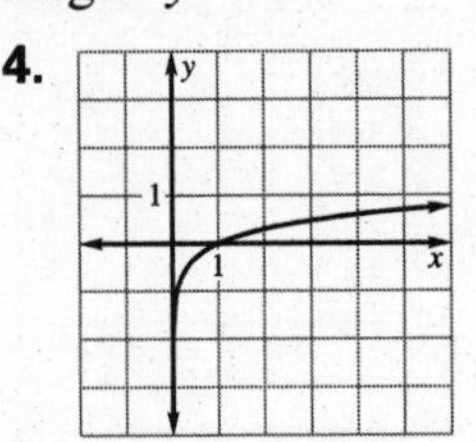

Domain: $x > 0$
Range: all real numbers

5.

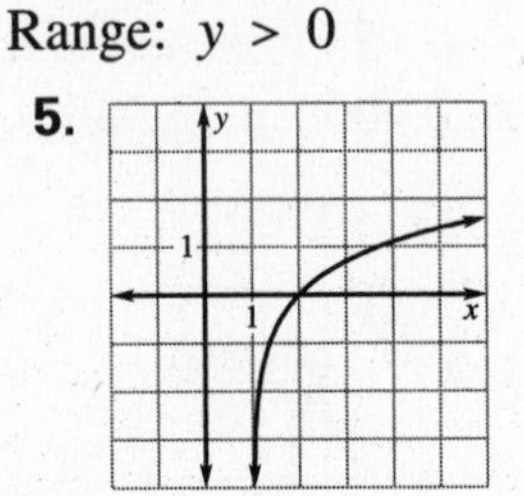

Domain: $x > 1$
Range: all real numbers

6.

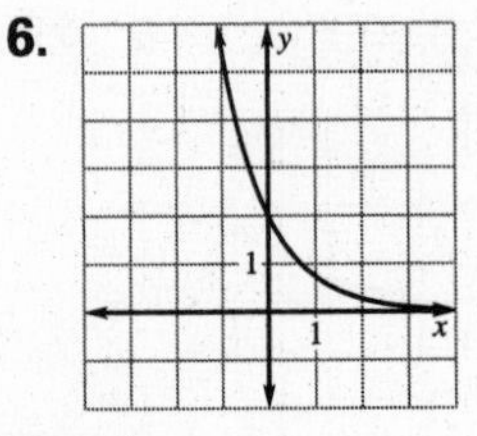

Domain: all real numbers
Range: $y > 0$

7. e^5 **8.** $\dfrac{3}{e}$ **9.** 4 **10.** 3 **11.** -3 **12.** -1
13. -2 **14.** 0 **15.** 1 **16.** -4 **17.** 5 **18.** 3
19. 4 (-8 is extraneous) **20.** exponential growth **21.** $y = 5^x$ **22.** 1.398 **23.** -1.079
24. $\log 7x^3$ **25.** $\ln 3 + \ln x + \ln y$ **26.** 1.431
27. $y = \frac{1}{4} \cdot 2^x$ **28.** $y = 2x^{1/2}$
29. $y = 20{,}000(.90)^t$; \$18,000 **30.** \$1127.50

Review and Assessment *continued*

Test B

Graph 1–6 on graph paper

1.

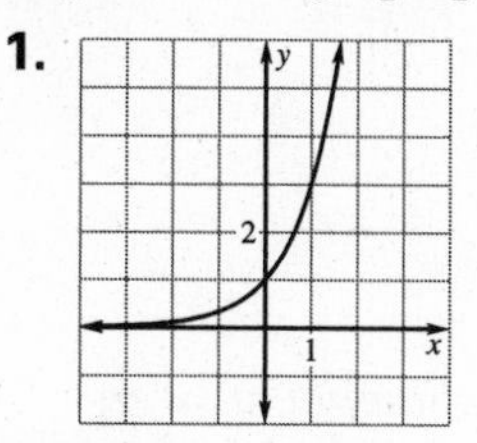

Domain:
all real numbers
Range: $y > 0$

2.

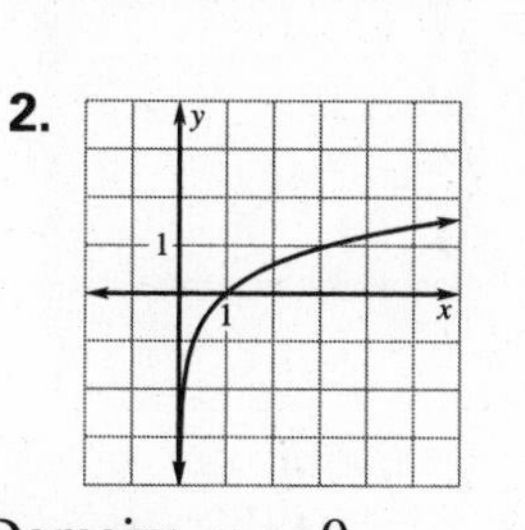

Domain: $x > 0$
Range:
all real numbers

3.

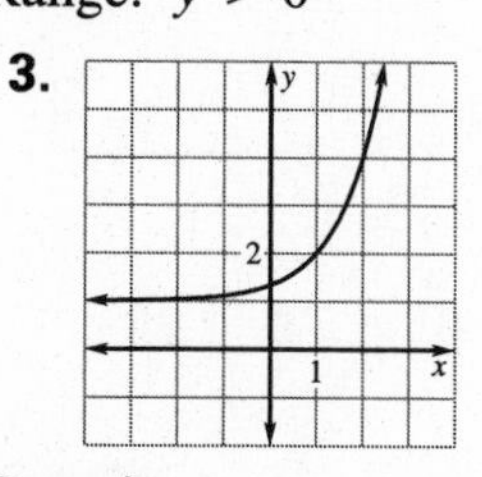

Domain:
all real numbers
Range: $y > 1$

4.

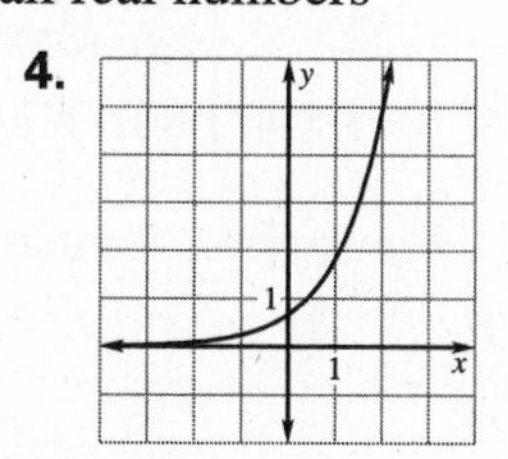

Domain:
all real numbers
Range: $y > 0$

5.

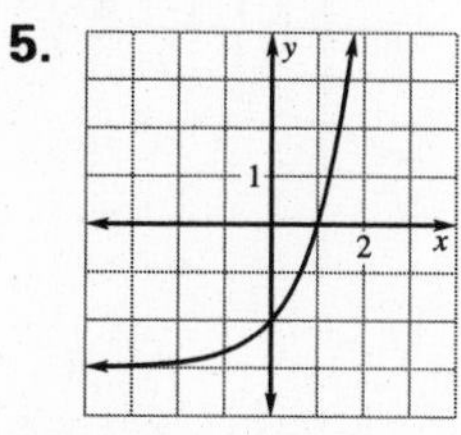

Domain:
all real numbers
Range: $y > -3$

6.

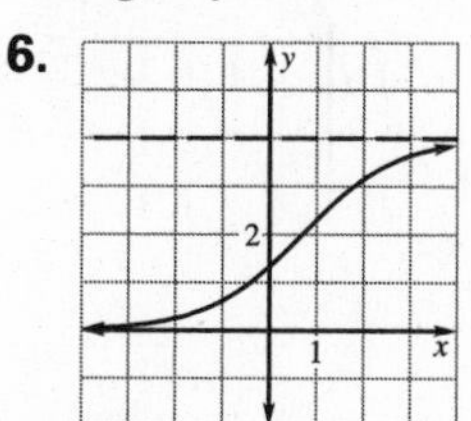

Domain:
all real numbers
Range: $0 < y < 4$

7. $-e$ **8.** $-\frac{8}{e}$ **9.** -2 **10.** 5 **11.** e^7

12. -4 **13.** -4 **14.** 0 **15.** 3 **16.** 16

17. 1 **18.** ln 5 **19.** 6 (-2 is extraneous)

20. exponential growth **21.** $y = 7^x$

22. 6.644 **23.** -0.903 **24.** $\log \frac{7}{b}$

25. $\ln 5 + \ln x - \ln 2$ **26.** 3.169

27. $y = 100\left(\frac{1}{2}\right)^x$ **28.** $y = 2.583x^{0.6309}$

29. $y = 18{,}000(.88)^t$; \$13,939 **30.** \$1161.83

Test C

Graph 1–6 on graph paper

1.

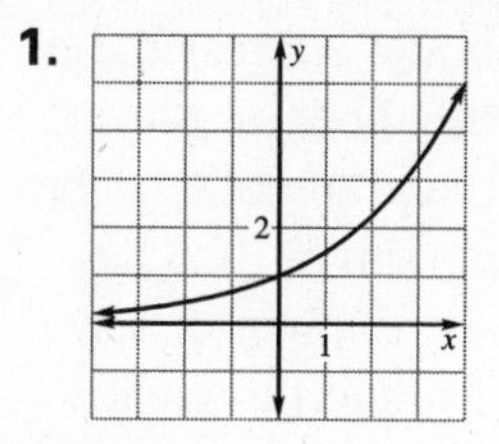

Domain:
all real numbers
Range: $y > 0$

2.

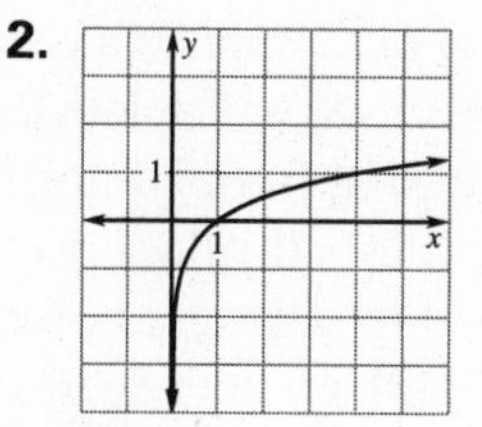

Domain: $x > 0$
Range:
all real numbers

3.

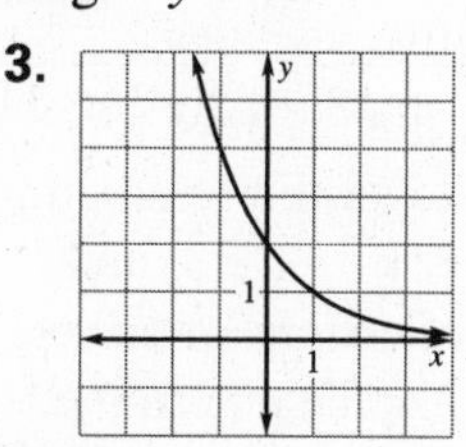

Domain:
all real numbers
Range: $y > 0$

4.

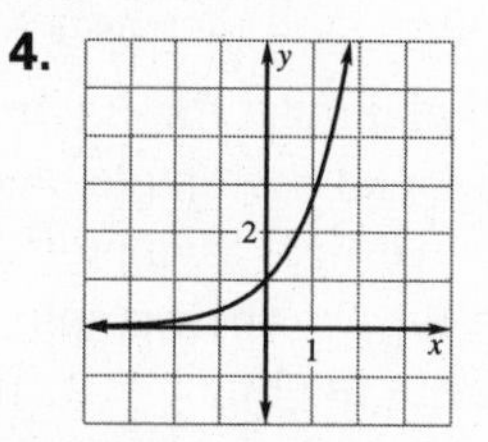

Domain:
all real numbers
Range: $y > 0$

5.

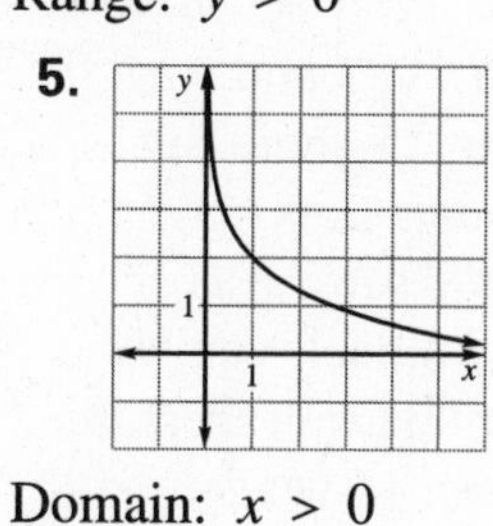

Domain: $x > 0$
Range:
all real numbers

6.

Domain:
all real numbers
Range: $y > 0$

7. e^4 **8.** $9e^2$ **9.** -3 **10.** 5 **11.** -2

12. -2 **13.** -3 **14.** 0 **15.** 2 **16.** 625

17. 2, -2 **18.** 8; (-4 is extraneous) **19.** 2

20. exponential decay **21.** $y = 8^x$ **22.** 4.428

23. 1.176 **24.** $\log_4 24$ **25.** $\ln 2 + \ln y - \ln x$

26. 2.481 **27.** $y = 3^x$ **28.** $y = 3.227x^{0.631}$

29. $y = 28{,}000(.92)^t$; \$18,454 **30.** \$1053.22

SAT/ACT Chapter Test

1. B **2.** C **3.** B **4.** D **5.** C **6.** C **7.** A

8. B **9.** C **10.** B

Review and Assessment *continued*

Alternative Assessment

1. a–c. Complete answers should address these points: **a.** • Explain that while both graphs are increasing from left to right, the graph of $g(x)$ has $y < 0$ for all values of x. The graph of $f(x)$ has $y > 0$ for all values of x. **b.** • Explain that $h(x)$ and $j(x)$ are both decreasing from left to right although $j(x)$ has $y > 0$ for all values of x, and $h(x)$ has $y < 0$ for all values of x. The logarithmic graph is increasing. It begins with negative y-values, and increases to positive y-values.

• Explain that the exponential graphs have a horizontal asymptote, and the logarithmic graph has a vertical asymptote. **c.** • $y = a \cdot 3^x$ and $y = \log_3 x$; Their graphs are reflections of each other in the line $y = x$. It can also be determined by rewriting the log equation in exponential form.

2. a. $P(t) = 50{,}000(1.051)^t$; 74,437
b. $P(t) = 100{,}000(0.919)^t$ **c.** 12,103; *Sample Answer:* The model began in 1980 and was good for 10 years. To estimate 25 years out is probably not accurate. **d.** *Sample Answer:* Town A has exponential growth, Town B is exponential decay. Exponential models can be used because there is a constant percent increase or decrease each year. **e.** about 14 years
f. about 3.4 years **3.** *Sample Answer:* 1985; One method would be to solve algebraically, another to solve graphically, or to solve using a table of values. Algebraically is more challenging since there are exponents with different bases on both sides.

Project: Investigating Zipf's Law

1. *Sample answer:* United States **2.** *Sample answer:* (1998 city population estimates)

City, State	*Population P*	*Rank R*
New York, NY	7,420,166	1
Los Angeles, CA	3,597,556	2
Chicago, IL	2,802,079	3
Houston, TX	1,786,691	4
Philadelphia, PA	1,436,287	5
San Diego, CA	1,220,666	6
Phoenix, AZ	1,198,064	7
San Antonio, TX	1,114,130	8
Dallas, TX	1,075,894	9
Detroit, MI	970,196	10

3. *Sample answer:* $y = -1.098x + 17.321$ (or $\ln R = -1.098 \ln P + 17.321$)

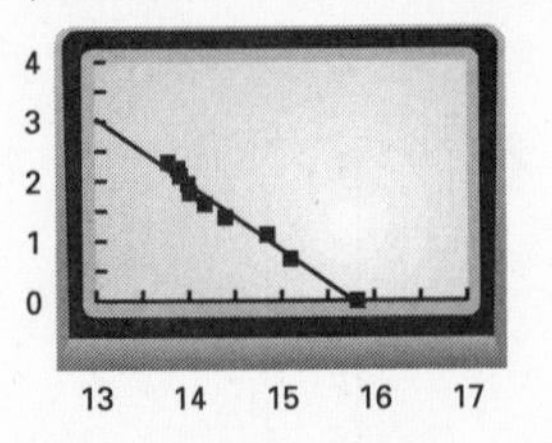

4. *Sample answer:* Yes **5.** *Sample answer:* $R = 33{,}300{,}000P^{-1.098}$ **6.** *Sample answer:* Population of San Francisco, California, is about 746,000. Model predicts a rank of 11.863; actual rank is 12. It matches very well.
7. Answers may vary.

Cumulative Review

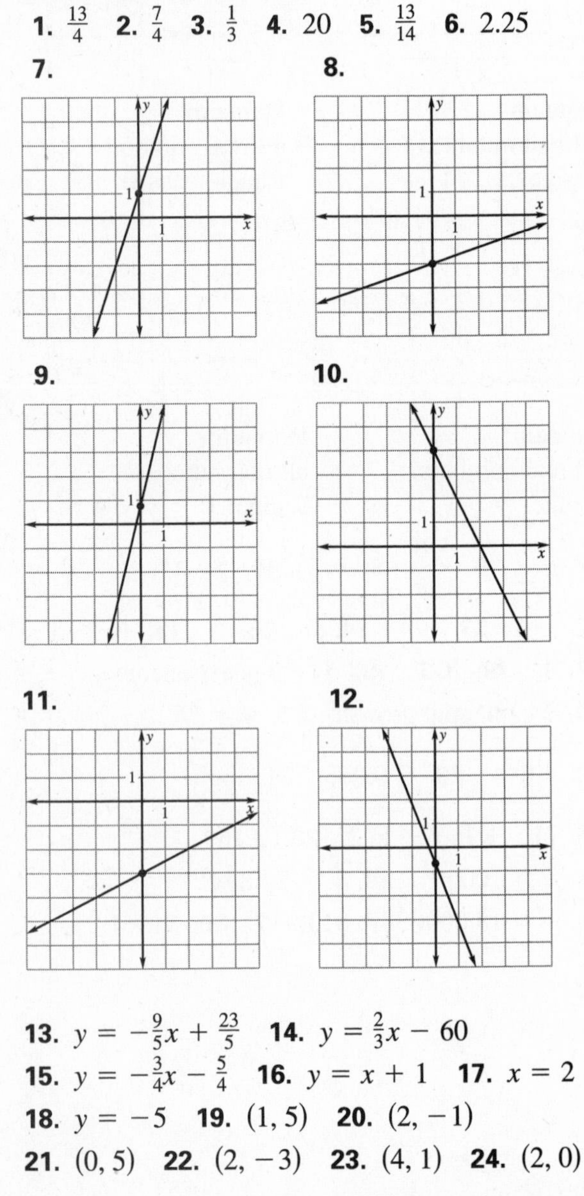

1. $\frac{13}{4}$ **2.** $\frac{7}{4}$ **3.** $\frac{1}{3}$ **4.** 20 **5.** $\frac{13}{14}$ **6.** 2.25
7. **8.** **9.** **10.** **11.** **12.**

13. $y = -\frac{9}{5}x + \frac{23}{5}$ **14.** $y = \frac{2}{3}x - 60$
15. $y = -\frac{3}{4}x - \frac{5}{4}$ **16.** $y = x + 1$ **17.** $x = 2$
18. $y = -5$ **19.** $(1, 5)$ **20.** $(2, -1)$
21. $(0, 5)$ **22.** $(2, -3)$ **23.** $(4, 1)$ **24.** $(2, 0)$

Review and Assessment *continued*

25. $(2, -1)$ **26.** $(-5, -3)$ **27.** $(2, 0)$
28. $(6, -5)$ **29.** $(12, 2)$ **30.** $\left(\frac{1}{2}, \frac{1}{3}\right)$
31. $y = x^2 - 7x + 12$
32. $y = -2x^2 - 13x - 15$
33. $y = -2x^2 - 12x - 19$
34. $y = -9x^2 + 18x - 4$
35. $y = 5x^2 - 20x + 22$
36. $y = 2x^2 - 6x + 5$
37. 4, 1 **38.** $-\frac{2}{3}$ **39.** 0, 2 **40.** $-11, -8$
41. $3, -3$ **42.** $4, -1$ **43.** $3\sqrt{3}$ **44.** $10\sqrt{2}$
45. $5\sqrt{2}$ **46.** $\frac{1}{3}$ **47.** $\frac{2\sqrt{5}}{5}$ **48.** $\frac{\sqrt{14}}{8}$
49. $-3 + 4i$ **50.** $45 - 11i$ **51.** $5 + 12i$
52. $\frac{16}{5} - \frac{8}{5}i$ **53.** $-\frac{3}{2} + 2i$ **54.** $1 - i$
55. $-2 \pm \sqrt{2}$ **56.** $\frac{5 \pm i\sqrt{3}}{2}$ **57.** $-3 \pm \sqrt{17}$
58. $\frac{-2 \pm \sqrt{6}}{2}$ **59.** $\frac{3 \pm \sqrt{15}}{3}$ **60.** $\frac{5}{2}, -\frac{1}{2}$
61. 0; one real solution **62.** 1; two real solutions **63.** -23; two imaginary solutions **64.** 160; two real solutions **65.** 0; one real solution **66.** 81; two real solutions **67.** $\frac{1}{25}$
68. $\frac{1}{216}$ **69.** 729 **70.** 27 **71.** $\frac{1}{729}$ **72.** $\frac{9}{4}$
73. $x + 2$ **74.** $x + 2$ **75.** $2x - 1 + \frac{5}{3x + 2}$
76. $x^2 + x - 3 + \frac{4}{2x - 1}$ **77.** $x + 3 + \frac{3x - 1}{x^2 + 3}$
78. $x^2 + 2x + \frac{5}{x^2 + 5}$ **79.** 25 **80.** 32
81. $\frac{17}{3}$ **82.** 125 **83.** 11 **84.** 8, 2
85. \$109,556.16 **86.** 1.9 years **87.** \$1176.43